Services and Business Process Reengineering

The book series aims at bringing together valuable and novel scientific contributions that address the critical issues of software services and business processes reengineering, providing innovative ideas, methodologies, technologies and platforms that have an impact in this diverse and fast-changing research community in academia and industry.

The areas to be covered are

- Service Design
- Deployment of Services on Cloud and Edge Computing Platform
- Web Services
- IoT Services
- Requirements Engineering for Software Services
- Privacy in Software Services
- Business Process Management
- Business Process Redesign
- Software Design and Process Autonomy
- Security as a Service
- IoT Services and Privacy
- Business Analytics and Autonomic Software Management
- Service Reengineering
- Business Applications and Service Planning
- Policy Based Software Development
- Software Analysis and Verification
- Enterprise Architecture

The series serves as a qualified repository for collecting and promoting state-of-the art research trends in the broad area of software services and business processes reengineering in the context of enterprise scenarios. The series will include monographs, edited volumes and selected proceedings.

More information about this series at http://www.springer.com/series/16135

Pradip Kumar Das · Hrudaya Kumar Tripathy ·
Shafiz Affendi Mohd Yusof
Editors

Privacy and Security Issues in Big Data

An Analytical View on Business Intelligence

Editors
Pradip Kumar Das
Department of Computer Science
and Engineering
Indian Institute of Technology Guwahati
Guwahati, India

Hrudaya Kumar Tripathy
School of Computer Engineering
KIIT University
Bhubaneswar, India

Shafiz Affendi Mohd Yusof
Faculty of Engineering and Information
Sciences
University of Wollongong
Dubai, United Arab Emirates

ISSN 2524-5503 ISSN 2524-5511 (electronic)
Services and Business Process Reengineering
ISBN 978-981-16-1009-7 ISBN 978-981-16-1007-3 (eBook)
https://doi.org/10.1007/978-981-16-1007-3

This Springer imprint is published by the registered company Springer Nature Singapore Pte Ltd.
The registered company address is: 152 Beach Road, #21-01/04 Gateway East, Singapore 189721, Singapore

To the God…,

To the Parents…

& To the Families…

Preface

Big data refers to collecting large volumes of data, giving us greater insight into our data which can be used to drive better business decisions and greater customer satisfaction. At this time, an increasing number of businesses are adopting big data environments. The time is ripe to make sure security concerns in these decisions and deployments, particularly since big data environments do not include comprehensive data protection capabilities, thereby represent low-hanging fruit for hackers. Securing big data is difficult not just because of the large amount of data it is handling, but also because of the continuous streaming of data, multiple types of data, and cloud-based data storage.

Primary purpose of this book is to provide insight about the security and privacy issues related to big data and its associated environmental applications. There are ten different chapters included in the study. Chapters 1 and 2 present a general discussion regarding various analytical issues concerning big data security. Different concerns and challenging factors are highlighted. Chapter 3 gives an insight about vulnerabilities of big data infrastructure and aims to alleviate fake data generation. Feature extraction with Cartesian moment functions is suggested to deal with fake data generation. Chapter 4 highlights the privacy threats, issues, and challenges of big data. Several techniques required to maintain data security have also been covered in brief. Chapter 5 deals with privacy concerns in big data databases. To address data misuse and privacy concerns, several anonymization techniques like K-anonymity, L-diversity, and T-Closeness anonymization methods are presented in detail and suggested to safeguard data privacy. Chapter 6 aims to highlight a succinct summary of frameworks to protect privacy and thereby address barriers to present big data-related architectures. It covers various big data-related polices and standards. Later, the Indian personal data protection bill is reviewed. Chapter 7 is concerned with data encryption and privacy preservation through multiple levels of encryption methods. Chapter 8 comprises mapping of benefits driven by big data analytics in healthcare domain. Later, the security and privacy concerns in healthcare sector are also addressed. Chapter 9 examine and elaborates the integration of big data and machine learning with cyber-security. Chapter 10 discusses the usage of big data and its related security concerns in business industry. Security threats that any business organization faces while working with huge amount of private data along with some counter measures to secure those data are thoroughly discussed here. In Chap. 11, governance of big data using data protection and privacy acts is discussed and ideas of deployment of these acts are noted. Few latest data security technologies in digital era are also highlighted.

Guwahati, India
Bhubaneswar, India
Dubai, United Arab Emirates

Dr. Pradip Kumar Das
Dr. Hrudaya Kumar Tripathy
Dr. Shafiz Affendi Mohd Yusof

Contents

About the Editors

Dr. Pradip Kumar Das is currently Professor in the Department of Computer Science and Engineering in IIT Guwahati. He completed his B.Sc. degree with Statistics major from Arya Vidyapeeth College, Guwahati, in 1989, and M.Sc. in Mathematical Statistics from Delhi University, North campus, and he was awarded the Ph.D. degree in Computer Science in the area of Automatic Computer Speech Recognition using Vector Quantization and Hidden Markov Modelling.

Dr. Das is a CSIR NET qualified JRF/SRF Fellow and worked in CEERI, Delhi, for 5 years and as Scientist Fellow in HRD group of CSIR (Automation section) for about two years. He has published more than 100 papers in international journals and conferences in India and abroad. Dr. Das has executed 14 sponsored projects and consultancies from agencies like MHRD, Department of Electronics, DST, Ministry of Social Justice and UNICEF. He has filed for a patent on speaker characterization. He has held the position of Organizing Vice Chairman, IIT JEE 2009, Vice President IIT Club, etc. He has visited numerous countries to present his research work in conferences and meetings. His research interests include speech recognition, analysis and characterization, image processing, Internet of things, AI, smart devices, algorithms and software engineering.

Dr. Hrudaya Kumar Tripathy completed his B.Tech. in Ceramics Technology from Indian Institute of Ceramics, Kolkata, MCA degree from Madurai Kamaraj University, and M.Tech. in Computer Science and Engineering from IIT Guwahati, and he was awarded the Ph.D. degree in Computer Science from Berhampur University.

Dr. Tripathy is currently an Associate Professor at the School of Computer Engineering, Kalinga Institute of Industrial Technology (KIIT), Deemed to be University (Institute of Eminence), Bhubaneswar, in India. He has 20 years of teaching experience in Computer Science at the undergraduate and postgraduate levels. Dr. Tripathy was invited as Visiting Senior Faculty by Asia Pacific University (APU), Kuala Lumpur, Malaysia, and Universiti Utara Malaysia, Sintok, Kedah, Malaysia. He was awarded the Young IT professional award 2013 on a regional level from the Computer Society of India (CSI). He has published many research papers in reputed international refereed journals and conferences. He is a senior member of

IEEE society, a member of IET, and a life member of CSI. Dr. Tripathy's research interests focus on machine learning, data analytics, robotics & artificial intelligence, speech processing & IoT.

Dr. Shafiz Affendi Mohd Yusof received the B.S. degree in Information Technology from University Utara Malaysia, Malaysia, in 1996, M.S. degree in Telecommunications and Network Management in 1998, M.Phil. degree in Information Transfer and Ph.D. degree in Information Science and Technology in 2005 from Syracuse University, Syracuse, USA.

He is currently Associate Professor at the Faculty of Engineering and Information Sciences, University of Wollongong in Dubai. He is Discipline Leader for Master of Information Technology Management (MITM) and Head of the Information Systems and Technology (INSTECH) Research Group. From 2012 to 2016, he was a faculty member of the School of Computing as Associate Professor in University Utara Malaysia. He held various other senior roles including Director of International Telecommunication Union—Universiti Utara Malaysia Asia Pacific Centre of Excellence (ITU-UUM ASP CoE) for Rural Information and Communication Technologies (ICT) Development and Deputy Director of Cooperative and Entrepreneurship Development Institute (CEDI). He is a certified professional trainer (Train of Trainers' Programme) under the Ministry of Human Resource, Malaysia, and has conducted several workshops on computers and ICT.

Chapter 1
Security in Big Data: A Succinct Survey

Akshat Bhaskar and Shafiz Affendi Mohd Yusof

1 Introduction

The term "big data" is defined by name itself as a large amount of data that is difficult or almost impossible to process using traditional methods. It can be any type of data that is found in our daily lives and is stored together as the most valuable assets in any organization, which can be used effectively and intelligently to give them support in decision-making, based on real facts instead of ideas. It is much faster, more reliable, and unique than any previous language, as well as it is faster and easier to manipulate data [1]. Big data is a combined term referring to large and complex data that is hard to handle and process by general software techniques like database management system, where data can be called as big data when either it is in collective amount so we can gain some pattern or knowledge from it or by analyzing it should give some value which can be useful. Using different big data technologies, patterns and knowledge can be developed such that it will be helpful in make better decision in critical areas such as machine learning, artificial intelligence, health care, economic production, predict natural disaster, etc.

The big data era has been bought with ample opportunities for scientific development, improving health care, economic growth, improving the education system, and various forms of entertainment [2]. The analysis of big data has to go through many stages to gain some meaningful value, which include some stages like data integration data acquisition, information cleaning, information extraction, query processing,

A. Bhaskar (✉)
School of Computer Engineering, Kalinga Institute of Industrial Technology (KIIT), Deemed to be University, Bhubaneswar, Odisha, India
e-mail: 1806098@kiit.ac.in

S. A. M. Yusof
Faculty of Engineering and Information Sciences, University of Wollongong, Dubai, UAE
e-mail: ShafizMohdYusof@uowdubai.ac.ae

P. K. Das et al. (eds.), *Privacy and Security Issues in Big Data*, Services and Business Process Reengineering, https://doi.org/10.1007/978-981-16-1007-3_1

data modeling, and interpretation. Every stage holds many challenges like heterogeneity, timeliness, complexity, security, and privacy of individuals [3]. One of the major issues in big data is security and privacy due to its huge infrastructure like large volume, velocity, and diversity. Although there are mainly four characteristics of big data security:

- Infrastructure and framework security
- Data privacy
- Data regulation
- Integral and reactive security.

The value of big data does not depend on how much data you have processed, but on what you are going to do with it. Data can be collected from many sources and later send to investigate and analyze further to find knowledge that allow lesser cost and time, new product expansion, prepared offerings, and intelligent decisions. With the help of big data and strong statistics, we can achieve many big organization-related tasks and concerns such as:

- Determining real-time failures, issues, and defects.
- Calculation risk portfolios
- Getting fraudulent behavior before it affects your organization.

Naveen Rishishwar and Tomar [4] in recent years, big data is comprised of five major Vs including which are also termed as characteristics of big data, as we can see in Fig. 1.

Volume: Big data name is defining this characteristic itself which is related to size. In general, volume refers to the hug amount.

Velocity: New data needs to be managed as well, so velocity defines the speed required to generate and processes data under appropriate time. In today's era, this can be easily done in real time with new technologies.

Value: Irrespective of how much data is available, it should must hold some meaningful value which can be useful for an organization otherwise it make no value. So the data must hold valuable information.

Variety: It is the different types of data which are collected for better calculations. This could be structured data or unstructured data as well.

Veracity: In simple word, it is the authenticity of the data. There will be no need to process those data for which you are not much confident that it will return some meaningful knowledge or not.

2 Big Data Security

While the big data snowball is speeding down the mountain of technical era to gain speed and volume, companies are trying to keep up with it. And they go downstairs, completely forgetting to put on masks, protective hats, gloves, and sometimes even skiing. Other than that, it is very easy to never cut it down by one piece. And putting all

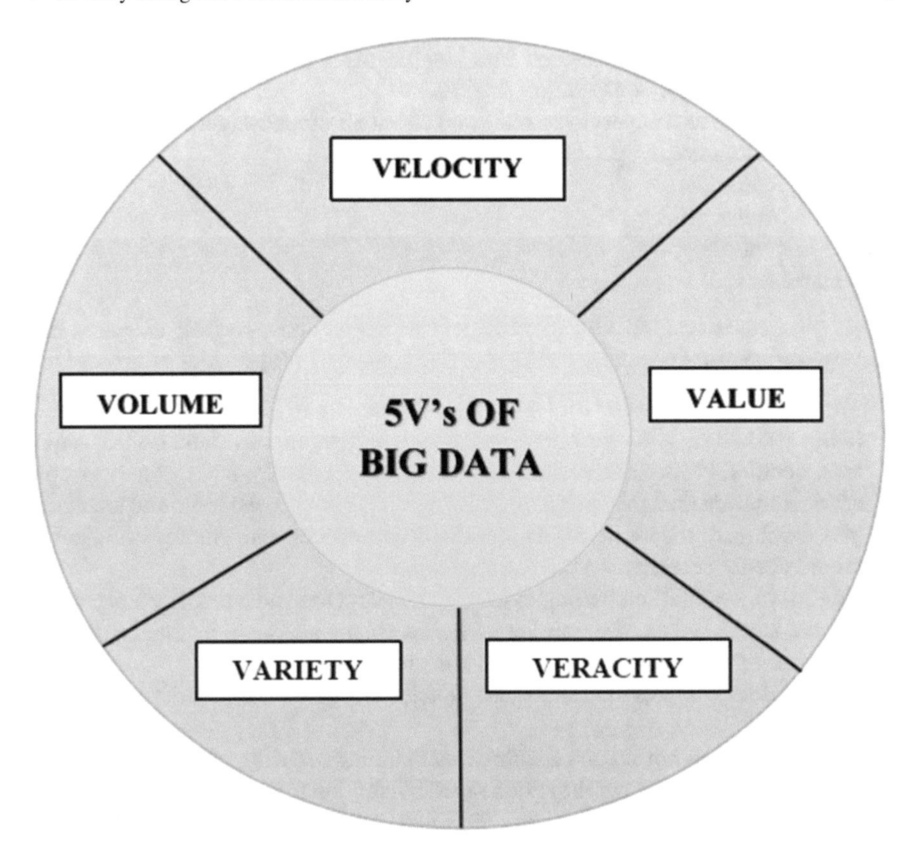

Fig. 1 Five Vs in big data

the precautionary measures at high speed can be too late or too difficult. Prioritizing low data security and putting everything up to the latest stages of big data acquisition projects could be a risky move. Big data security is defined by all the tools and technologies required to monitor any kind of attack, theft attempt, or other security breaches. Like every other cyber-security attack, big data can be compromised from online or offline domains. These threats include the theft of individual data or an entire organization. There could be indirect attacks as well like DDoS attack which can crash the server. During big data analysis, the private information of individuals collected by social networks or feedback needs to be merged with huge data sets to find meaningful patterns; sometimes, unintentionally in the whole process, confidential fact about a person might become open to the world. Often, it lead to privacy risk and violation of privacy rights. Some hackers or thieves who know better about big data take advantage of those who do not know much about this technology. Some big data technical issues and challenges are:

1. Processes need to be divided into smaller tasks and allocate these tasks to different node for computation purpose.
2. Treat a node as a supervising node and check all other assigned nodes to see if they are functioning properly.
3. Fault tolerance.
4. Data heterogeneity
5. Data quality
6. Scalability.

As big data is continuously growing in size day by day, so their concerns like security and privacy preservation are also rising regularly [5]

- The major reason for security and privacy concerns in big data is that now it is easily available and accessible to everyone. It has become so common that scientists, doctors, business executives, government employees, and ordinary people are also sharing data on a large-scale every day. However, the tools and technologies developed to date to handle these vast volumes of data are not sufficient to provide better security and privacy to the data.
- Nowadays, available technologies are not sufficient to handle security and privacy threats, and they lack the training as well as many adequate features and basic fundamentals to secure these vast amounts of data.
- Big data does not have much adequate policies that guarantee security and privacy measures.
- Technologies are not much capable of maintaining security and privacy, leading to many cases daily where they get tampered intentionally or accidentally. Thus, it is required to improve current algorithms and approaches to prevent data leakage.
- There is a lack of funding in the security sector by a company to protect their crucial data. It turns out that a company should spend at least 10% of its IT budget on its security but on average, less than 9% is being spent, making it harder for itself to protect its data.

Kaur and Kaur [6] some important security and privacy concerns related to big data are as follows:

- Secure data storage and transaction logs.
- Security practices for non-relational data stores
- Secure computations in distributed programming frameworks
- End point input validation/filtering
- Real-time security monitoring
- Scalable and composable privacy-preserving data mining and analytics
- Cryptographically enforced data-centric security
- Granular audits.

3 Background Study

Various important functions in this domain are performed. Some important and relevant inventions are discussed in this section. Thuraisingham [7] unveils a comprehensive overview of big data and its privacy and security. Sharif et al. [8] discussed Verizon (a service-based security) embedded security model to protect its cloud. It has split security infrastructure into two major parts, one for the authority and the other for the data center domain. Parmar et al. [9] proposed encryption of data at rest in the proposed Hadoop encryption system used for encryption and decryption but it has been observed limitation to the fact that the MapReduce functions reduces its performance. Fugkeaw et al. [10] proposed that it focuses on expanding the access control framework called the Collaborative Cipher Policy Attribute Role-based Encryption (C-CP-ARBE) to provide better control over large data extensions in the cloud. Li et al. [11] proposed an algorithmic calculation of knowledge arrangement to balance load on technologies and later improve accessibility and accountability. Zheng and Jiang [12] introduced a stand-alone conference that joins the Kerberos conference engineering and SAML implementation [13, 14]. Other data sources that are not organized into logs, images, audio, and video files, etc., have no predefined feature where some more data sources are emails, XML, CSV, TSV files, etc. [15].

4 Solution to security in Big Data

There are so many threats which are challenging our technology to secure big data as every second there could be a very big loss and it can lead to great risk or failure. Keeping all of this in mind, we have some of the general practices which can help us in preventing data better than feel sorry. Here, we are going to see two most common practices which are following:-

Access Control and Internal Security: Threats are not always from outside of organization; it can be anyone or it can be internal part as well either could be employee knowingly or unknowingly data can be compromised by them also. Such that accessing any big data frameworks either Hadoop on any cloud technologies by anyone should be taken seriously. While most employees do not try to leak information to a private company, there are many ways they can do it unknowingly. Companies should take care of the recruitment, evaluation, and evaluation of potential employees with sensitive information in the workplace. In addition, establishing and communicating security policies in advance and reviewing safety standards through training is always a necessary step in improving data security among employees. Once employees are hired and trained, organizations should deploy infrastructure security. It can be done by following some general practices on infrastructure security like authentication security, data monitoring, maintaining integrity in files within system, user activity monitoring and by deploying data-centric security.

Endpoint validation: Users should ensure that the source of data is not malicious neither it should be transferred without deploying encryption like cryptography methods and if it is, then it should filter malicious input materials generated by that source. Well it could be more better with the idea of "bring your own device" model as it reduces the risk by good measure. We can use some techniques like trusted certificates, proximity-based approach, statistical similarity detection technique, outlier detection techniques, antivirus or malware protection, and many more to validate endpoint inputs.

Data Encryption: It can be a good solution to big data security issues. Possible kinds of encryption of data and information are the following:

File system-level encryption: It is mainly used to protect the sensitive information and file or folder level inside the tools like Hadoop or the cloud itself. This is not as reliable as it can be compromised when running within the system as it can decrypt at the operating system level.

Database encryption: The main idea is to encrypt the whole database which can be performed along with file system encryption and there are multiple techniques available for this like transparent data encryption and column-level encryption.

Transport-level encryption: This encryption is used to protect data from getting lost or tampered while moving from one end to another. It can be done with the help of SSL/TLS protocols.

Application-level encryption: This method uses APIs to protect data at the application side of the user by access control from any kind of invalid authentication.

Storage-level encryption within Hadoop: This level of encryption is deployed within Hadoop and it mainly came into the role when there are chances of physical theft or loss of entire disk volume. This option uses transparent data encryption within the Hadoop distributive file system (HDFS) to make a safe landing. Although this method can slow down the system.

5 Analysis of Different Privacy-Preserving Techniques

Technologies do not really come with only benefits; along with this, they came with many scientific problems and challenges. Big data has a slightly brighter future than other data science technology in IT sector, which also leads to more responsibility [16]. In last few decades, many algorithms and techniques have been developed by humans and machines to provide better security and privacy [17]. Here, we are going to analyze some traditional techniques which have been used for decades and still there in implementation.

Major traditional techniques:

- Data perturbation
- Data encryption
- Data anonymization.

Table 1 Comparison of privacy-preserving techniques

Techniques	Major advantage	Major disadvantage	Sub-techniques
Data perturbation technique	Installation cost is less along with easy implementation	Algorithms are not same for every data, which also lead to more complexity	Random perturbation, randomized response, blocking, differential privacy protection
Data encryption technique	Directly applied on data, no data breaches, high protection	Complex due to different encryption keys for data, Compatibility issues and maintenance expenses are high	Watermarking key, data anonymity algorithm, data provenance technology, access control techniques, etc.
Data anonymization technique	Easy implementation, real-world applicable and cost effective	Risk of data defects and unintentional data leak also data pull off could be more anonymous	Data masking, de-identification: • K-anonymity • L-diversity • T-closeness

By Table 1, we can see different techniques applicable to secure big data along with their pros and cons. While implementing techniques on particular requirements, it is concluded from above tables that noisy and distortion-based techniques are more reliable. To provide perfect protection in big data domain, we can think of encryption technique but this technique has high cost and high communication overhead as we can see in Table 2. While on the other hand, we have data perturbation technique which computing cost and communication overhead is low but it may lead to high data loss which make him unacceptable; on the other hand, we have data anonymization technique which provides more balanced environment and has reliable algorithm, frameworks, and properties like K-anonymity L-diversity, etc. It can protect privacy at lower cost comparing with encryption technology and lower data loss compared to perturbation technique.

Table 2 Performance analysis

Techniques	Preservation – scale	Data dependency	Data loss	Computing cost	Communication overhead
Data perturbation	Medium	High	High	Low	Low
Data encryption	High	Low	Low	High	High
Data anonymization	High	Low	Medium	Medium	Low

- Data anonymization technique analysis

It is the technique which provides privacy by hiding the valuable data and user information. De-identification is traditional technique which implements the concept of anonymization. At first, raw data is categorized into sensitive data by data mining, publishing, etc. Whereas to achieve privacy, some de-identification operation and methods get applied such as generalization, suppression, decomposition, interference etc. before releasing it to further processing. Generalization is mainly used to hide the user's identification, whereas suppression is to not release data at any cost. Decomposition is to mix and shuffle the attributes and interference is exchanging and modifying data by adding noise to the data. Due to large amount of data, low data loss could also be much information for attackers to practice re-identification. For example while logging in fraud application via Facebook, attackers collect the sensitive content which get posted by personal or community on their feed or profile and target those users for wrong intentions with the help of data collected through their profile.

Currently, there are three methods in de-identification:

- K-anonymity
- L-diversity
- T-closeness (Table 3)

Table 3 De-identification privacy-preserving methods analysis and computational complexities in big data

Preserving methods	Definitions	Limitations	Complexity
K-anonymity	It is a framework which provides privacy by releasing only those attributes which can be revealed and hide left sensitive attributes. For say there is k attributes then it releases k-1 attributes such that they match other attributes	Homogeneity-attack, background knowledge	O(k logk)
L-diversity	It is a group-based anonymization technique used to provide privacy by decreasing granularity of data representation	It is not easy to achieve and implement. And insufficient to prevent attribute disclosure	$O((n^2)/k)$
T-closeness	It is defined as distribution of sensitive attribute in every group/class such that it does not cross a threshold distance from attributed distributed in the whole table	It requires distribution in manner such that sensitive attributes should not cross the threshold	2O(n)O(m)

6 Big Data Security in Agriculture and Farming

Agriculture is not just a profession, it is a way of survival for farmers as well as consumers. Farming is playing a major role in the growth of food production worldwide. It is a knowledgeable skill that has been passed down from generation to generation.

Sahoo et al. [18] with the increasing population, the demand for food is also increasing day by day. In one of the United State govt organization surveys, it is stated that by 2050, the world's population will exceed 10 billion. This can lead to an increase in demand for food by at least 40% in today's scenarios. To meet the demand, the agricultural sector needs to increase production capacity by 1.5 times. As can be seen in Fig. 2, the shortage section in graph is showing scarcity which can be possible if production growth will be same.

Whereas in the world full of technologies and innovation, it is also possible to provide some solutions which can help in growth of food production. Without any doubt, this maybe groundless to accept some decades ago but in today's era, we can use technologies like IoT, artificial intelligence, data mining, and big data to provide insights and better knowledge to farmers which can be very useful for them

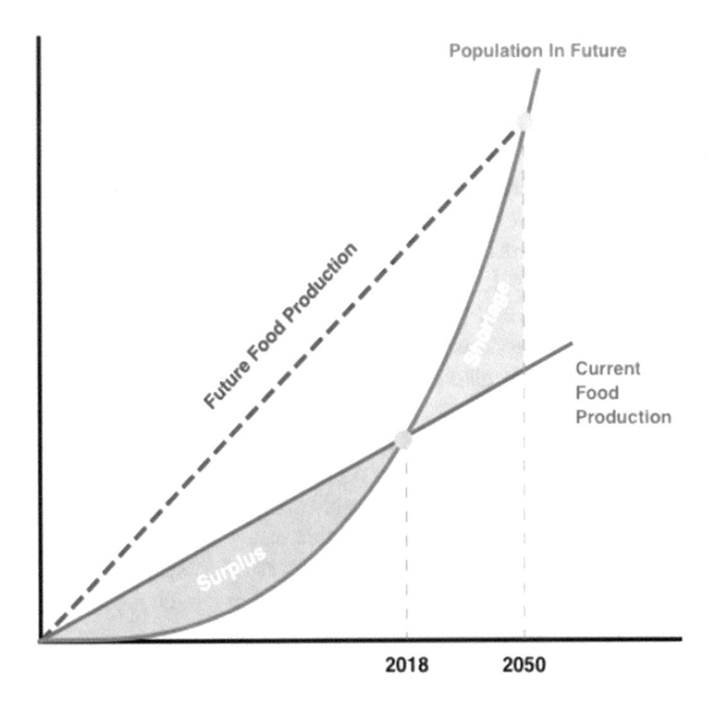

Fig. 2 Required food production in the future as compared to current production

to produce more amount of food in same resources and reduce wastage. For example, weather detection and climate prediction could give better idea about when to grow and how to protect farming.

- ***How Big data can Boost Agriculture Growth?***

To grow farming, the major goal in agriculture is to reduce food wastage as much as possible. There are various techniques that are helping such as smart irrigation, smart equipments, weather prediction, humidity detector, and many more intelligent infrastructure.

Big data is considered as a combination of technologies which help in collecting data of different environment and process it furthermore to find valuable pattern which assists better decision-making. For instance, if some particular type of grass is not suitable for some environment, farmers can deploy techniques to control it through insights of data collected through various sensors and machines.

Big data could help in following ways by collecting insights data:

- Development of new seed traits by mapping collected data to access plant genome
- By analyzing of crop health, seed quality, and drought conditions.
- Food tracking such as using sensor to collect data about moisture and humidity, and it also prevent from spreading of crop-borne illness.
- Improving better supply chain of seeds, fertilizers, equipment, etc. to the farmers.

Mishra et al. [19] big data technology is in very prior stage of implementation, although in near future, it is going to be most required technology for farming as predicted by experts. Now let us have look on some of use cases of big data.

- ***Top use cases for big data on the agriculture and farming:***
  - *Managing environmental challenges*: Climate changes are the major threat to the farming as it can even lead to crop waste such as drought conditions or heavy rainfall. Data-driven farming can help to make it easier through regular monitoring of climate by enabling intelligent resources and machines.
  - *Using pesticides ethically*: With precision farming, farmers can monitor the heath condition of crop and what kind of pesticides should be used when and by how much. It can also helpful for government to examine and provide chemical less fertilizers and pesticides for long-term health of crop.
  - *Farm equipments*: Many IT and agriculture companies are working on equipment kit so that farming can be more better by deploying these equipments which include sensors, cloud-based real-time data, climate detectors and many more. It can be life savior for farmers that can help in making better decision by collection of data through this machines as it let farmers have idea of what is the condition of crop health or humidity or how much resources are need to grow it faster and better.
  - *Supply chain challenges*: It has been seen that there is very much gap between supply and demand which is creating problems for farmers as well as buyers. Either it is about equipment supply to framers or food supply to consumers,

there is a big scam of dealers as well which cause poverty in farmers irrespective of agriculture being one of the most prominent and necessary profession to survive. Also, consumers are getting food at higher price than the normal price.

In order to supply and reduce market needs, big data can help in achieving supply chain efficiently by tracking the food and improving delivery routes. It will not only make farmer smarter but also more productive, efficient, and intelligent.

- **Role of Big Data Security in Agriculture supply chain:**

Agriculture supply chain is a general term for supplying food from suppliers to distributors. In order to do that, there are many problems and threats that could arise; one of the threats is related to security. Here, security is not just limited to the theft or loss of food but it is also about data security that has been collected from the supply chain.

Data in the supply chain is collected from various methods which include wireless sensor data, communication from warehouse and transport, RFID, GPS location, vehicle position, shipment tracking, public communication such as call recordings, container tracking, tracker equipments, black boxes in airplanes and heavy vehicles, and many more.

Now, in order to efficient and good use of data, there is a framework for big data security in agriculture supply chain as shown in Fig. 3.

This system proposes framework that is designed to provide security. Below are the objectives that can be achieved:-

- IoT enabled system which helps in tracking of agriculture goods through WSN and GPRS
- Deploy data science software components with techniques like data mining and data extraction to analyze and find pattern from incoming big data stream in this domain. Along with this, lightweight annotations can be deploy to find and solve data pollution and data noise.
- Intrusion detection system (IDS) could be deployed to maintain security not only for incoming big data but also stored data from intruders and thefts.
- Advance techniques and methods for analyzing and processing big data. It can help in real-time visualization of extracted knowledge and to identify valuable data only from all available data.
- The use of less expensive farming technology as well use the program in the real project to improve the efficiency of feeding the agriculture business by reducing the cost of food damaged due to poor storage and shortage of grains supply chain.

- **Proposed Big Data Security Framework:**

In major part of the world, supply chain of food is still dependent on many traditional ways like bar code scanning physical data collection, and transfer of important information from one source to another. As a result, many problem occurs such as

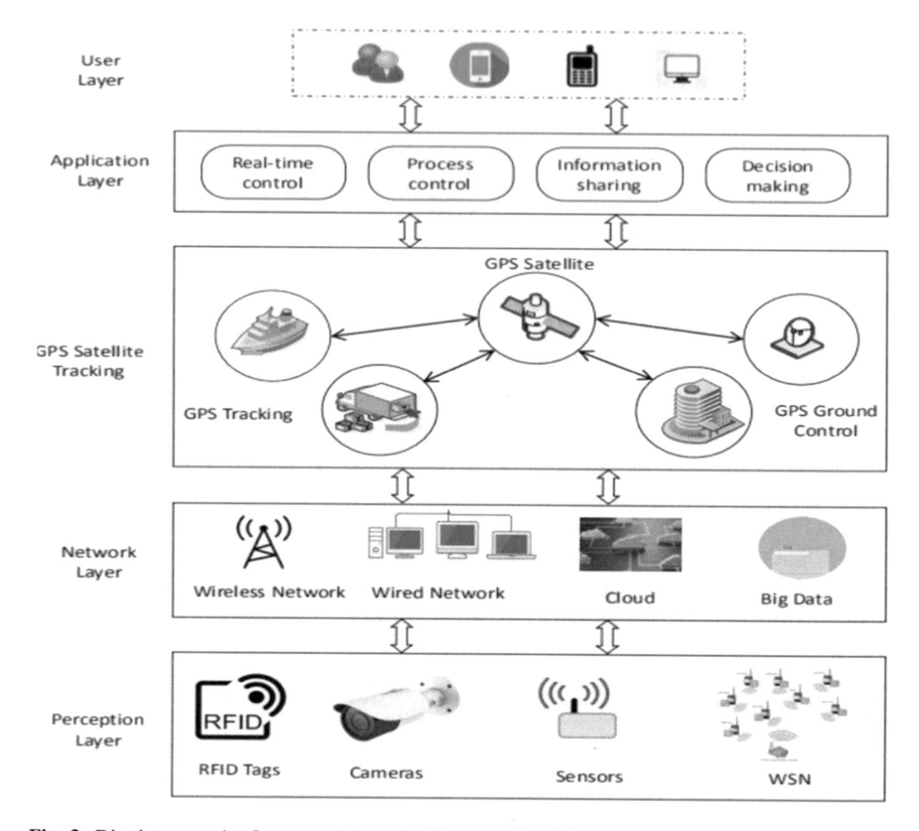

Fig. 3 Big data security framework in agriculture supply chain

delay delivery, error, and miss communication. These problems can be solved with the help of Internet of things and big data that can build and intelligent system which can be more accurate and error free system.

In this section, we will see the proposed framework for deploying big data security framework in the whole agriculture supple chain. This system is designed with intention of providing better services to all the corners of supply chain including government, distributor, and farmers as well. In general, data get collected from various sources in real time and put into logistic network which lead to large amount of big data. This framework will capture incoming data store them into secure area and extract valuable data from it in order to improve the efficiency of supply chain and reduce load on the system. Proposed framework is traced through several stages as shown in Fig. 4.

This framework is mainly divided into following four sections:

- Big data aggregation: Its main function is to collect data from various sources such as IoT enabled sensors, WSN, GPRS, camera, etc.

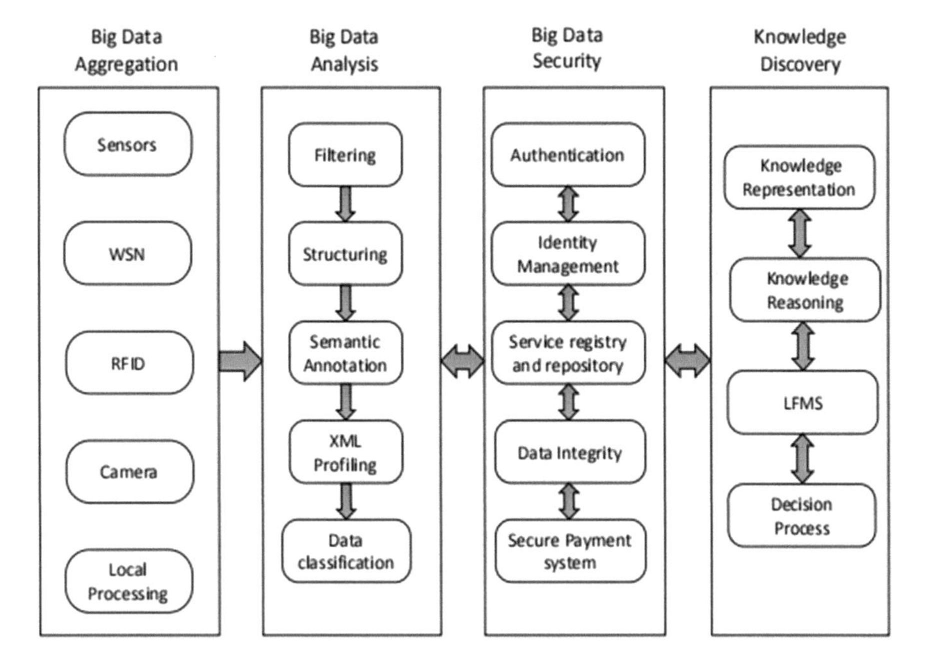

Fig. 4 Framework for integrating big data security into agriculture supply chain

- Big data analysis: This section deals with analyzing of collected data through various processes like data cleaning and noise removing etc. Based on the different types of data such as structured and unstructured data, semantic annotation is applied on the data. During the annotation on text various methods like extraction, identification and association of data is applied. When annotation get completed, all the annotated data is stored into XML file which further goes for data classification.
- Big data security: This section is dedicated to big data security in agriculture. Security is more needed functionality to ensure trust and reliability of usages. Agriculture in one of those field who does not backed up with advanced devices such that security is also not provided to every region of agriculture. Currently, there is not any good mechanism has developed to provide security to IoT in farming. Such that mechanism is needed to ensure trust of security and privacy. Therefore, systems like lightweight intrusion detection system are needed to deploy in IoT field of farming which ensure no harm to data stored. It should provide basic security functionality like authenticity verification, identity management, data integration, repository guarding, and to provide better secure payment. Farms must take responsibility of securing data which is stored on cloud as well.
- Knowledge discovery: This is the section which deals to resolve, aggregate, and find interesting pattern automatically from big data through various expert knowledge system methodologies such as knowledge representation and reasoning. LFMS known as local farm management system is used to manage and utilize

farms in efficient manner with the help of data which is collected through its interfaces and furthermore it could be beneficial in limiting the conflicts of experts and enhancing decision-making procedure.

7 Conclusion

Big data is a leading revolutionary mechanisms in storing data. A variety of big data is being collected on a daily basis that cannot be ignored. As the data is growing and becoming very efficient in the decision-making process, along with a better future in every organization and field, at the same time, it is also coming up with new threats and security challenges. In this paper, we tried to summarize some common and basic security issues which can not be ignored where for privacy preserving and security solution, some techniques can be used like monitoring, filtering and encryption. These are quite good, but every method or algorithm has some pros and cons so that more new algorithms are needed to revise these techniques over time, as well as increase speed and accuracy. The era of big data has just begun and this technology has to go much further. More problems will occur and more solutions will be required. Therefore, further research is needed to develop a streamlined and robust system.

References

1. Mishra M, Mishra S, Mishra BK, Choudhury P (2017) Analysis of power aware protocols and standards for critical E-health applications. In: Internet of things and big data technologies for next generation healthcare. Springer, Cham, pp 281–305
2. Mishra S, Mishra BK, Tripathy HK, Dutta A (2020) Analysis of the role and scope of big data analytics with IoT in health care domain. In: Handbook of data science approaches for biomedical engineering, Academic Press, pp 1–23
3. Mishra S, Tripathy HK, Mishra BK, Sahoo S (2018) Usage and analysis of big data in E-health domain. In: Big data management and the internet of things for improved health systems, IGI Global, pp 230–242
4. Naveen Rishishwar V, Tomar K (2017) Big data: security issues and challenges. Int J Tech Res Appl 42(AMBALIKA): 21–25. e-ISSN: 2320-8163
5. Pathrabe TV (2017) Survey on security issues of growing technology: big data. In: IJIRST, National Conference on Latest Trends in Networking and Cyber Security, March 2017
6. Kaur G, Kaur M (2015) Review paper on big data using hadoop. Int J Comput Eng Technol 6(12):65–71
7. Thuraisingham B (2014) Big data—security with privacy. NSF Workshop, September 16–17
8. Sharif A, Cooney S, Gong S, Vitek D (2015) Current security threats and prevention measures relating to cloud services, Hadoop concurrent processing and big data. In: 2015 IEEE international conference on big data (Big Data). IEEE, pp 1865–1870
9. Parmar R, Roy S, Bhattacharaya D, Bandyopadhyay S, Kim TH (2017) Large scale encryption in hadoop environment: challenges and solutions. IEEE Access
10. Fugkeaw S, Sato H (2015) Privacy-preserving access control model for big data cloud. In: 2015 International computer science and engineering conference (ICSEC). IEEE, pp 1–6

11. Li P et al (2016) Privacy-preserving access to big data in the cloud. IEEE Cloud Comput 3(5):34–42
12. Zheng K, Jiang W (2014) A token authentication solution for hadoop based on kerberos pre-authentication. In: 2014 international conference on data science and advanced analytics (DSAA). IEEE, pp 354–360
13. Mishra S, Mahanty C, Dash S, Mishra BK (2019) Implementation of BFS-NB hybrid model in intrusion detection system. In: Recent developments in machine learning and data analytics. Springer, Singapore, pp 167–175
14. Mishra S, Sahoo S, Mishra BK (2019) Addressing security issues and standards in Internet of things. In: Emerging trends and applications in cognitive computing. IGI Global, pp 224–257
15. Bertino E (2015) Big data—security and privacy. In: 2015 IEEE international congress on big data, New York City, NY, USA, June 27 - July 2, 2015 pp 757–761
16. Mishra S, Tripathy HK, Mallick PK, Bhoi AK, Barsocchi P (2020) EAGA-MLP—an enhanced and adaptive hybrid classification model for diabetes diagnosis. Sensors 20(14):4036
17. Jena L, Kamila NK, Mishra S (2014) Privacy preserving distributed data mining with evolutionary computing. In: Proceedings of the international conference on frontiers of intelligent computing: theory and applications (FICTA) 2013. Springer, Cham, pp 259–267
18. Sahoo S, Mishra S, Panda B, Jena N (2016) Building a new model for feature optimization in agricultural sectors. In: 2016 3rd international conference on computing for sustainable global development (INDIACom), New Delhi, 2016, pp 2337–2341
19. Mishra S, Mallick PK, Jena L, Chae GS (2020) Optimization of skewed data using sampling-based preprocessing approach. Front in Publ Health 8:274. https://doi.org/10.3389/fpubh.2020.00274

Chapter 2
Big Data-Driven Privacy and Security Issues and Challenges

Selvakumar Samuel, Kesava Pillai Rajadorai, and Vazeerudeen Abdul Hameed

1 Introduction

Data is everywhere and in many forms. Basically, a large complex and varied amount of data from a domain or sector called as Big Data. Big Data is a diamond mine for industry, business, and service sectors of this century. [1] Data analytics and business intelligence tools, techniques, methods, and technologies help the process of analyzing this Big Data for finding hidden patterns, correlations, and creating insights for strategic decisions [2].

Most of the data being collected and stored in private organizations when we want to use software applications, devices such as communication devices, tools, and information technologies. This data shall be shared with third-party organizations [3]. That brings several privacy and security risks. The explosion of devices which are interconnected and to the Internet, the amount of data accumulated and processed is growing day after day, which poses new issues and challenges related to privacy and security [4, 5]. The main known reason for this issue is the lack of standards and regulations.

This chapter mainly will serve as the introductory chapter for this book and introduce the weaknesses and the areas could be improved, particularly in Big Data-driven privacy. Generally, more research works and solutions are available for data security but not much focus on the privacy issues and challenges, particularly individual's privacy matters are not much focused. Therefore, more Big Data-driven privacy research and solutions are required.

S. Samuel (✉) · K. P. Rajadorai · V. A. Hameed
Asia Pacific University of Technology and Innovation, Kuala Lumpur, Malaysia
e-mail: selvakumar@staffemail.apu.edu.my

K. P. Rajadorai
e-mail: kesava@staffemail.apu.edu.my

V. A. Hameed
e-mail: vazeer@staffemail.apu.edu.my

P. K. Das et al. (eds.), *Privacy and Security Issues in Big Data*, Services and Business Process Reengineering, https://doi.org/10.1007/978-981-16-1007-3_2

2 Big Data and Their Characteristics

Data and its characteristics have been evolving tremendously in this era [6]. Basically, the data and its management can be categorized as multiple generations. The database management systems (DBMS)/relational database management systems (RDBMS) age can be considered as first-generation data types, the business intelligence system (BIS) with the data warehouse age is the second-generation data types, and the Big Data analytics age is the third-generation data types.

Big Data is applied to datasets which cannot be able to manage by first and second-generation database management software tools and techniques to capture, store, access, and analyze the data. Big Data has been created from various Internet of things (IoT) devices, machines, gadgets, appliances, equipment's, smartphones, software applications, software systems, banking systems, e-payment systems, email system, and many other sources [7].

Table 1 illustrates eleven Big Data characteristics such as volume, velocity, variety, veracity, validity, volatility, value, variability, visualizations, valence, and vulnerability [8]. The Big Data properties bring big security and privacy issues and challenges due to technical deficiencies, organizational culture, and environmental factors [9].

3 Big Data-Driven Security

Security alludes to the methods, strategies, and technical measures used to forestall unapproved get to, change, stealing of information, or physical harm to gadgets and systems (Sun Z et al. 2018). The Big Data security concerns are same as other data types—to protect its privacy, trustworthiness, and availability [17].

Table 1 Big Data characteristics and their concepts [10]

V's	Characteristics	Concepts
1	Volume	The first important property is volume, which refers to the amount of data being accumulated
2	Velocity	The second most important property is velocity, which refers to the data flow rate into the organizational memory
3	Variety	The third important property is variety, which refers to various forms and types of data being collected
4	Veracity	The fourth important property is veracity, which refers to the trust worthiness, availability, and quality of the data being collected
5	Validity	The fifth property is validity, which is related to Veracity and it refers to the applicability of data in a context [11]

(continued)

Table 1 (continued)

V's	Characteristics	Concepts
6	Volatility	The sixth property is volatility, which is related to temporal aspects of the data and it determines how long it is valid to maintain in the organizational memory [12]
7	Value	The seventh important property is value, which refers to the value add to the respective organizations/businesses through insights created from the data being collected
8	Variability	The eighth property is variability, which refers to inconsistencies in which variable data sources could load data into the data storage in variable speeds, formats, or types [13]
9	Visualizations	The ninth property is visualizations, which refers to different ways of data representation such as dashboards, heat maps, cone trees, and k-means clustering to improve data insights [14]
10	Valence	The tenth property is valence, which refers to the interrelationships between the collected massive data. If interconnections between the data is established, they can add value to the organization [15]
11	Vulnerability	The eleventh most important property is vulnerability, which relates to the security, privacy, and technology risks in data being captured [16]

Table 1 represents the Big Data-driven security issues and challenges dependent on the properties of Big Data. These difficulties directly affect the structure of security settings that are required to handle every one of these properties and requirements [18] (Table 2).

Cloud Security Alliance (CSA) has organized the Big Data security challenges into three types such as integrity and reactive security, data management, and infrastructure security [25]. This will become four if we include the data privacy. The infrastructure security refers to the security of data storage, computations, and the other infrastructure of a data center. The data management security challenge refers to the secured data provenance, access, and other aspects of the data management. Lastly, the integrity and reactive security refers to security aspects such as real-time observation of inconsistencies and attacks [26]. Additional details which are related to the points above are discussed with the respective sub-titles in the following section.

4 Some Imperative Security Issues and Challenges

Some important security challenges created by Big Data are discussed here. The volume of opportunities present by Big Data is lesser than the challenges and issues generated. The common solution for this is encrypting everything to make data secure nevertheless, wherever the data is stored [27]. Basically, the available solutions address the general data security issues and measures; no most reliable solutions are available to overcome the Big Data-driven security issues and challenges.

Table 2 Big Data-driven security challenges or issues based on their characteristics

Big Data Characteristics	Security challenges or issues created by Big Data
Volume	Support to a major number of intruders [19]. Therefore, big security measures are required
Velocity	Physical security risks [20]. Produce outline of the person's snap and position [21]. Therefore, the data protection risk is high
Variety	Numerous organizations have not appropriately safeguarded and protected the semi-structured and unstructured data [22]. Therefore, protection mechanism which is equivalent to structured data is required for the unstructured data as well
Veracity	Security penetrate identified with a major number of charge cards. It shows the weakness in the current security solutions
Validity	Data leakage is a common problem due to improper management of data. Therefore, it shows the weakness in the current security mechanisms or management
Volatility	This issue is like data validity. Most of the companies particularly small size organizations do not maintain the individual's data after a certain period due to storage limitations and expenses associated with maintaining the data. Therefore, the unattempt data may be a threat and challenge for Big Data companies
Value	Big Data creates big value to the respective organizations. Therefore, more appropriate security mechanism should be applied to manage this challenge
Variability	It can also refer to anomaly detection that can benefit the organization [23] and all the above seven V's could be affected by the eighth dimension of Big Data, namely variability. Therefore, a new security mechanism is required
Visualizations	Security policies related to the visuals from various tools should be established in addition to assigning access controls and privileges based on user roles and responsibilities [24]
Valence	Security management procedures should maintain the level of performance for both current and future development of Big Data eco-system
Vulnerability	The vulnerabilities of sensitive data leakage must be identified and appropriate measures to review the confidentiality, integrity, and availability of Big Data systems and data are required. Therefore, the data security may be ensured

The security mechanisms being used for first and second-generation data base management systems have unsuccessful to adapt to the versatility, interoperability, and flexibility of contemporary advances that are required for Big Data [28]. Moreover, traditional encryption and anonymization of data are not sufficient to overcome the Big Data issues and challenges. They are sufficient to secure static data but are not adequate when information computation is engaged, as data computation is common in Big Data platforms. The current mechanism to prevent the data using security controls is weak. A new approach is required to prevent an attacker access the data in case an attacker violates the security controls which is placed at the edge of the networks [29].

The HP's Open Web Application Security Project (OWASP) (OWASP 2014; Jose A et al. 2015) has identified some important security issues such as insufficient security in mobile, web and cloud interfaces, insufficient authorization, insecure network-related services, insufficient data transfer encryption, privacy issues, insufficient physical security and security configurations, and security configurations, and firmware. This clearly reveal that the current security mechanisms are insufficient to manage the Big Data eco-system.

4.1 AI Applications and Big Data Security

The development of artificial intelligence (AI) applications, and Big Data domain has been facing many new and unknown security challenges. AI methods such as machine learning and deep learning have been helping to expand the application of Big Data in all core industries and service sectors immensely.

The machine learning and deep learning applications can identify the vulnerable Big Data management automatically, nevertheless on the other hand AI applications will be able to collect the data automatically as well. This brings very complicate security challenges and privacy issues in the Big Data management.

4.2 Fake Data Generation and Fake Mappers

Fake data can be generated by the cybercriminals if they have managed to access the organizational data and store it into a data lake. The crucial challenge here is that the organizations are unable to identify the fake data which is stored in the data lake. In case we generate an analysis report from this, data may get a false report, resulting in a serious loss of revenue. This challenge can be solved at some extents by applying several fraud detection methods [30] (Fig. 1).

4.3 Fake Mappers

In Big Data engineering, next to the data gathering, the collected data undergoes parallel processing using one of the methods called MapReduce. At this point, the data will be divided into several parts. Then a mapper processes them and designates everything to storage preferences. The current security settings could be altered in case an intruder has managed to access your mappers' code, or it can be replaced with fake mappers. This will produce the faulty MapReduce process, whereby intruders

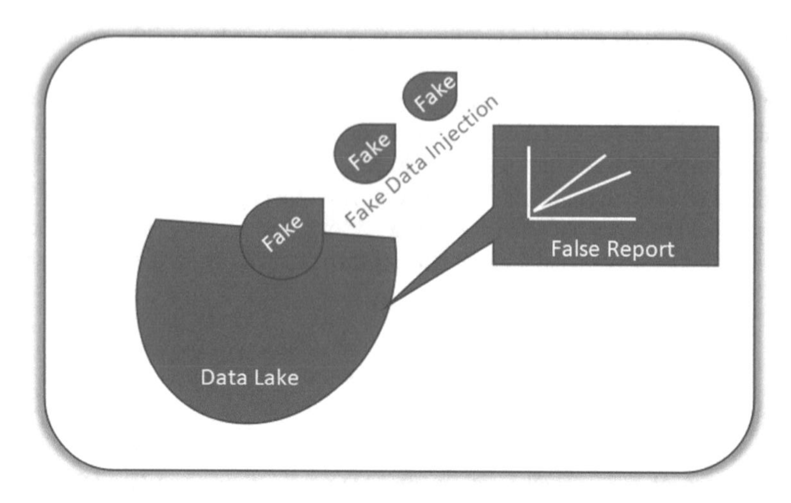

Fig. 1 Fake data generation

can be benefited. This challenge is due to insufficient protection provided by the Big Data domain [31].

4.4 *Granular Access Control*

One of the essential functional elements in Big Data environment to provide access rights for users is granular access control. This access control restricts the access of certain data in a data set, even a user needs access to other parts of the data. This leads to obscures maintenance and performance of the Big Data system. In a Big Data environment, it is difficult to grant access to all parts of the data in case a user really need to access it, for instance, to conduct a sensitive research on the data because the Big Data technologies themselves were not designed this way. Furthermore, this access control can become more challenging after the use of increasingly large data sets and complex dashboards. Eventually, this will open more vulnerabilities and it may take more time to find a breach in the Big Data environments.

4.5 *Data Provenance*

Data provenance is a record that portrays entities and procedures involved in creating and conveying that data resource [32]. It is very useful to determine the origin of a breach, as this method can be used to track the flow of data using metadata. However, there are pitfalls and risks in maintaining the data provenance [33].

Data provenance is a substantial Big Data issue. This concern is not new, but it is an ongoing issue. It is critical in security point of view. Because, unauthorized modification in metadata will produce the wrong data sets, this can make it difficult to find the information you need. Apart from unauthorized changes, program code also modifies data, which will create additional opportunities to make it difficult to maintain data provenance. Furthermore, undetectable data sources can be a major barrier to tracing security breaches and cases of fake data generation.

4.6 Real-Time Big Data Analytics Security Concerns

Real-time Big Data analytics is referred as analyzing large volume of data as soon as it enters the system. A major challenge faced in real-time analysis is the ambiguous definition of real-time and the random requirements that result from different interpretations of the term. As a result, businesses must invest considerable time and effort to gather specific and comprehensive requirements from all stakeholders to adopt a specific definition of real-time, what data sources should be used for it. Then the next difficulty is creating a capable architecture. In addition, the architecture must have the ability to handle rapid changes in data size and be able to measure it as the data grows. Implementing security solution for these analytics is complicate and produces a large volume of data of its own accord. Software solutions should be designed to prevent prompting misleading alarms of violation alerts when there are no real threats. This false alert can divert from the real risks of attack.

However, on the other hand, real-time analytics can be used to provide solution for real-time streaming security concerns. Authors in [34] explored that real—time security examination which can help observing streams continuously and identify and reduce these attacks. By utilizing these analytics, clog can be promptly identified and illuminated as fast as could reasonably be expected. This is the positive side of the real-time Big Data analytics.

The other aspect of streaming Big Data concerns in terms of IoT is discussed in the following section.

4.7 Big Data and IoT Security Concerns

IoT sensor devices are the main source for streaming data. An infinite flow of streaming data coming from the various sensor devices and instruments, for instance, stock price data, heath care data, etc. IoT-based data is yet another important dimension in the Big Data domain.

The IoT-based Big Data infrastructure brings new type of privacy and security concerns. Authors in [35] did an analysis on existing IoT solutions and determined that 70% of them have security and privacy issues. These issues mostly related to authorization, encryption, firm ware, data mobility, and strategies.

The connectivity between physical devices and networks in an IoT application causes Big Data security is a crucial issue as it can even damage the smart devices that are deployed if the security is vulnerable to attacks. A lot of IoT services are built by utilizing the endpoint devices and platform that is equivalent to communications, computing, and IT solutions. Endpoint devices including the Internet for computer hardware devices in TCP/IP networks such as a computer, laptop, mobile, tablets, printers, smart meters, etc. Besides, the configuration of low complexity devices, as well as rich devices and gateway that link the physical and digital worlds, is considered as an endpoint. These endpoint devices are an additional source for Big Data and IoT security concerns. Big Data is important to get the defensive services in IoT security as it compiles an abundant volume of data from each smart object or endpoint devices that generates a large stream of data over time.

4.8 Cloud and Big Data Security Concerns

Currently, the cloud infrastructure is the main storage option for Big Data. The marriage between Big Data and cloud storage raised many security issues such as data Loss, malicious insiders and data breaches due to trust, loss of control over data, and multi-tenancy issues. These issues and challenges are not new; however, these issues are big now due to Big Data. Therefore, most of the major cloud vendors' service-level agreements (SLAs) are not guaranteed the required levels of security and privacy, particularly for their consumers [36]. The following sections briefly described the three major concerns such as trust issues, loss of control over data, and multi-tenancy.

4.8.1 Trust Issue

Confidence on cloud providers performs a key role in capturing clients by reassuring cloud service vendors. Because of the loss of control over data (discussed in Sect. 4.8.2 below), consumers have relied on trust on cloud resources as an alternative. Consequently, cloud service vendors develop trust among their consumers, and their operations are certified in accordance with company's safety measures and regulations.

4.8.2 Loss of Control Over Big Data

Loss of control over data is one more security breach that can occur where the cloud provider hosts consumer data, applications, and resources on its premises. Since consumers have no control over their data, cloud service vendors can process their consumer's data, which will cause privacy and security concerns. Moreover, cloud vendors back up data on different storage locations, it is not possible to guarantee that their data will be eliminated all over in case consumers remove their data. This issue can lead to abuse of the undeleted data. In this case, users see the cloud service vendors as enigmatic as they cannot track their data resources transparently.

4.8.3 Multi-tenancy

Multi-tenancy implies that the sharing of physical and virtualized resources between numerous consumers. Utilizing this setup, an assailant might be on same computer as the target. Cloud service providers apply multi-tenancy characteristics to create scalable infrastructure that can effectively meet the needs of customers. However, sharing resources multi-tenancy implies that the attacker can easily access the target data. This is an important security challenge when Big Data stored in a cloud infrastructure.

4.9 Summary

In summary, to alleviate the Big Data security challenges and issues in an organization, three points can be considered. The first point is to ensure the data security, the organization should come up with a balanced approach toward policies, regulations, and analytics with the help of best practices, whereby organizations can handle massive data and perform useful analytics without compromising the performance and adequate security, secondly should secure the infrastructure with the technologies which have the adequate security protections. There are technologies such as MapReduce, Storm, Hadoop, Mahout, Hive, Piglatin and Cassandra do not have adequate security protections, and thirdly should secure the access methods and indexing and query processing using reliable data management practices, which includes the data integration policy and ensure the quality of data.

5 Big Data-Driven Privacy

Privacy is a state in which one is not observed or disturbed by other people and free from public attention. But, the privacy of individuals is being hacked by some of the Big Data companies. Privacy is an individual’s right. These organizations have been following all the data of people either in public or in private. In most of the cases, the individual does not aware of this. An individual is at risk even with worldwide Big Data organizations, because the accessible solutions are not intended to secure the consumers privacy in the Big Data era.

Table 3 outlines the Big Data-driven privacy dependent on the attributes of Big Data. These difficulties directly affect the structure of privacy measures that are required to handle all these characteristics and necessities.

Table 3 Big Data-driven privacy challenges or issues based on its characteristics

Characteristics of Big Data	Big Data-driven privacy challenges or issues
Volume	Create huge value, data is influence, and data is wealth. Therefore, the individual's privacy is being sliced by the companies
Velocity	Capture the real-time location data and personal details. Therefore, the individual's privacy is not considered by most of the companies
Variety	Cannot viably oversee information containing delicate data. Therefore, the individual's privacy is more vulnerable
Veracity	Time variation data of people is concern identified with privacy. Therefore, the individual's privacy is more concerned
Validity	Data leakage is a common problem due to improper management of data. Therefore, the individual's privacy is a big question
Volatility	This is like data validity. Most of the companies particularly small size organizations do not maintain the individual's data after a certain period due to storage limitations and expenses associated with maintaining the data. Therefore, the unattempt data may be a threat for individuals
Value	Big Data creates big value to the respective organizations. Therefore, more breach for the individual's privacy and companies finding more ways to capture the data
Variability	It can also refer to anomaly detection that can benefit the organization and all the above seven V's could be affected by the eighth dimension of Big Data, namely variability. Therefore, the individual's privacy is not a specific concern here
Visualizations	Privacy policies related to the visuals from various tools should be established in addition to assigning access controls and privileges based on user roles and responsibilities
Valence	Privacy management procedures should maintain the level of performance for both current and future development of Big Data systems
Vulnerability	The vulnerabilities of sensitive data leakage must be identified and appropriate measures to review the confidentiality, integrity, and availability of Big Data systems and data are required. Therefore, the individual's privacy may be ensured

A few of major Big Data organizations can control and access the greater part of the individual's data of the world's total populace and practically all the information on the Web. This is perhaps the greatest hazard to privacy. When an individual want to download an application or a game, he or she should agree with the companies to access the individual's mobile device cameras, locations, etc. although which are totally not relevant to the service provided.

Once the sufficient data captured from the consumers, the companies can disconnect from the communication network, whereby the companies can minimize the security risk, but this is one of the biggest risk to the individual's privacy due to the data stored in the Big Data companies. Ensuring the individuals privacy is the responsibility of data-driven companies as they are extremely benefited by creating values from the captured data or by selling the data to the third party.

5.1 Some Good Measures

There are some measures, regulations, standards, and approaches are available to treat the individual's privacy protection is much reasonable. The healthcare sector relatively provides better protection for the individual's privacy such as the Health Insurance Portability and Accountability Act (HIPPA) than the other sectors.

The European countries practicing a user-friendly data collection model called as opt-in approach to protect the individual's privacy is stricter. That is, the European nations do not permit organizations to utilize individually recognizable data without the person's earlier assent. The organizations must illuminate the people when they gather data about them and reveal how it will be stored and handled. This is an opt-in approach.

The confidentiality and fair use are the two key factors of privacy. To protect the confidentiality, the privacy-enhancing technologies and systems can be used to enable users to encrypt email, conceal their IP address to avoid tracking by web server, hide their geographic location when using mobile phones, use anonymous credentials, make untraceable database queries, and publish documents anonymously. There are numerous applications use completely homomorphic encryption which permits encrypted inquiries on database, which keeps secret private consumer data where the data is regularly stored. An investigation has proposed privacy extensions to UML to help software developers rapidly envision privacy requirements and program them into Big Data applications.

5.2 Challenges and Recommendations

The solutions available to protect the individual's privacy are not sufficient. More detailed and more stringent standards, policies, regulations, and approaches are required. Basically, the Big Data companies are collecting data from their service or application users globally. However, all the available standards are either country or continent based. Therefore, the global-level standards, approaches, regulations, and policies are required to overcome this issue.

The USA practicing a data assortment model called opt-out approach, which allows companies to collect data and use it for other marketing purposes without acquiring the permission from the person whose data is being gathered and afterward utilized. This approach makes the most people in a generally impeded position. The companies and countries are better to adopt the opt-in approach to duly honor the individual's privacy.

Nevertheless, the healthcare sectors have a better standard and approaches to protect the individual's privacy, but lacking details. A study on patient information privacy and security demonstrated that 94% of hospitals had in any event one security penetrate in the previous two years. In most cases, the attacks were from an insider instead of outer.

The clinical services sector is recording the information in electronic clinical records and pictures, which is utilized for transient well-being checking and continual epidemiological exploration programs. There are no clearly informed procedures given to capture and store the data.

The privacy is a person's entitlement to control the data collection, utilizes, or dis-terminations of their recognizable data. But, most of the individuals does not aware of this. With the consideration of this, a simple privacy aware data collection model is suggested for a basic healthcare application which is collecting data from the patients and providing healthcare advice to the patients is depicted in Fig. 2.

This model is just illustrating a sample information collected by an application not all. Likewise, to protect the individual's privacy, a more detailed and stricter approaches are required for any purpose, or in any occasion, the data is being collected by any sectors. Every single data item acquired from the individual's patients should get prior consent from the respective individuals with the reason for the requested data item. Every time when an individual data is accessed, an alert message should send to the respective individuals with the reason for access. This may make the companies/healthcare sector in a disadvantage position, but this kind of approach will ensure the individuals privacy in a better way.

6 Research in the Big Data-Driven Privacy and Security

The researchers have published a research work in 2018 based on Big Data research literature published in SCOPUS from 2012 to 2016. They have downloaded and examined 13,029 scripts titles, abstracts, and keywords published for the period of 2011–2016 in journals. The research result reveals that among the major Big Data areas published in journals, only 360 (2.1%) articles were published on privacy and security topics. This shows that less focus has been given to Big Data-driven privacy and security research even though the Big Data research is quickening at an exponential rate from 2011.

Most of the research works not showing the in-depth analysis of security and privacy issues. Particularly, solutions for the privacy issues are not focused. The reason for this might be the privacy issues mostly related to the individuals, not to the Big Data companies. Table 4 from the same research work of evidently shows that the human and societal aspects of security and privacy are the less focused research area. Particularly, this aspect of research supposed to be focused on individual's privacy matters. This clearly shows that the weaknesses or ignorance of the individual's privacy-related research works.

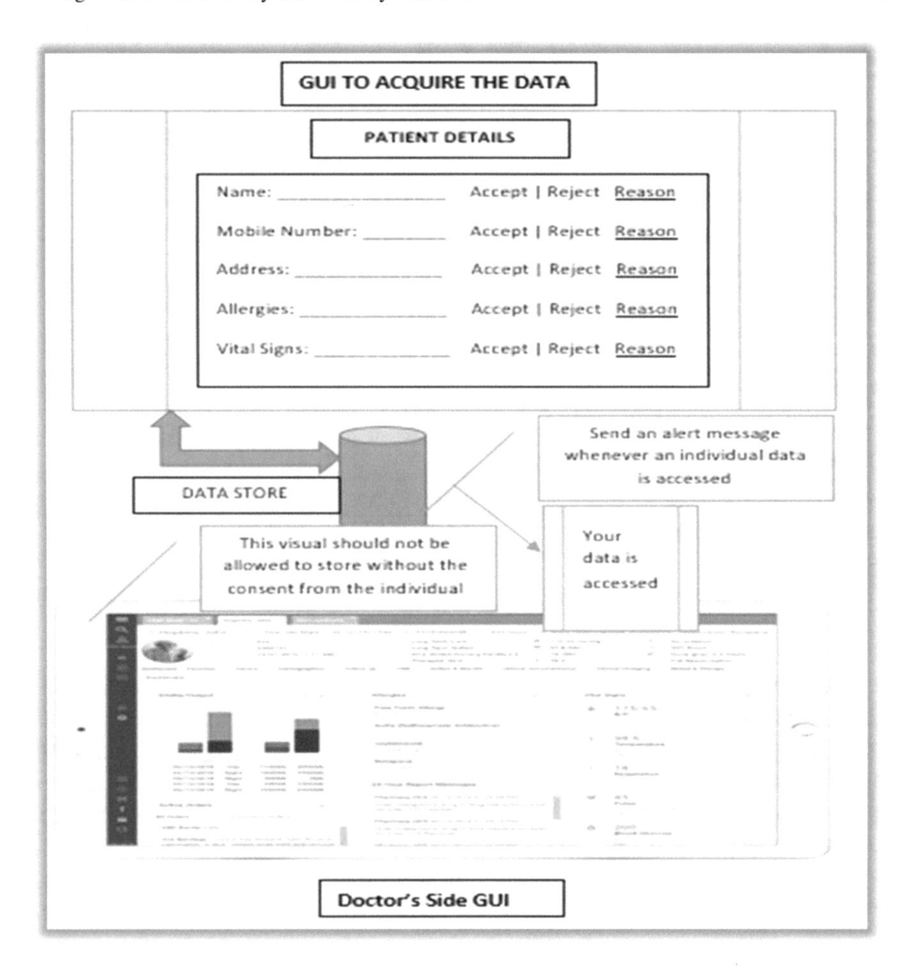

Fig. 2 Approach to acquire, store, and access an individual data

6.1 *Some Good Concerns Related to Big Data-Driven Privacy and Security Research*

As mentioned in Sect. 3, the core Big Data security objectives such as to preserve its confidentiality, integrity, and availability have not much different with any other data types. Hence, the data security research have been conducted by the researchers for first and second-generation data types are still applicable to Big Data-driven security research, but as discussed in Sect. 4, all the available traditional solutions should be re considered in terms of Big Data characteristics and challenges. Apart from that, the AI-driven challenges and issues are also considered.

Be that as it may, this is not the situation for Big Data-driven privacy research. Most of the research works on security and privacy have proposed solutions mainly

Table 4 Research in privacy, security and Big Data based on Scopus searched data [31]

Levels 1 and 2 in privacy and security	2012 No.	2013 No.	2014 No.	2015 No.	2016 No.
Cryptography	31	44	81	144	226
Formal methods and theory of security	0	0	2	1	1
Security services	49	74	158	260	372
Intrusion/anomaly detection and malware mitigation	3	3	12	11	11
Security hardware	10	23	25	42	73
System security	103	182	331	583	818
Network security	64	89	209	363	551
Database and storage security	10	7	22	14	33
Software and application security	12	24	37	64	60
Human and societal aspects of security and privacy	0	0	0	1	2
Securoty and privacy	40	70	147	237	318

to secure the organizational data. It makes the most individual users of the Big Data company services and applications relatively in a disadvantaged position as not much focus on the data privacy. To resolve the privacy issues, privacy-driven research needs to be conducted to secure individual's privacy. As a result of the research global-level updated laws, policies, regulations, and norms should be built up, not the nation / continent based, or organization based.

In light of the above concerns, the following points can be considered when a researcher want to conduct a Big Data-driven security and privacy research:

1. Data being used in the research should possess all the necessary characteristics of Big Data; otherwise, the results produced by the research may not be reliable for Big Data-driven security and privacy issues and challenges.
2. To maintain the consistency in the research, individual's privacy matters should be considered equally with the Big Data-driven security and challenge matters.
3. Global policies, standards, approaches, and regulations should be developed with the consideration of the local standards of all Big Data companies.

References

1. Agarwal N, Tripathi A (2015) Big data security and privacy issues: a review. Int J Innov Comput Sci Eng 2:12–15
2. Almeida F (2018) Big data: concept, potentialities and vulnerabilities. Emerging Sci J. https://doi.org/10.28991/esj-2018-01123

3. Bekker A (2020) Big data security: issues, challenges, concerns [online] Scnsoft.com. Available at: https://www.scnsoft.com/blog/big-data-security-challenges. Accessed 21 July 2020
4. Mishra S, Mallick PK, Tripathy HK, Bhoi AK, González-Briones A (2020) Performance evaluation of a proposed machine learning model for chronic disease datasets using an integrated attribute evaluator and an improved decision tree classifier. Appl Sci 10(22):8137
5. Mishra S, Tripathy HK, Mallick PK, Bhoi AK, Barsocchi P (2020) EAGA-MLP—an enhanced and adaptive hybrid classification model for diabetes diagnosis. Sensors 20(14):4036
6. Chen M, Mao S, Liu Y (2014) Big data: a survey. Mobile Netw Appl 19:171–209. https://doi.org/10.1007/s11036-013-0489-0
7. Coppens S, Gil Y et al (2010) Provenance XG final report [online] W3.org. Available at: https://www.w3.org/2005/Incubator/prov/XGR-prov-20101214. Accessed 20 July 2020
8. Rath M, Mishra S (2020) Security approaches in machine learning for satellite communication. In: Machine learning and data mining in aerospace technology, pp 189–204. Springer, Cham
9. Mishra S, Sahoo S, Mishra BK (2019) Addressing security issues and standards in Internet of things. In: Emerging trends and applications in cognitive computing, pp 224–257. IGI Global
10. Rath M, Mishra S (2019) Advanced-level security in network and real-time applications using machine learning approaches. In: Machine learning and cognitive science applications in cyber security, pp 84–104. IGI Global
11. Demchenko Y, Ngo C, de Laat C, Membrey P, Gordijenko D (2014). Big security for big data: addressing security challenges for the big data infrastructure. In Jonker W, Petković M (eds) Secure data management, pp 76–94. Springer International Publishing. https://doi.org/10.1007/978-3-319-06811-4_13
12. Dudkin I (2018) Top 5 big data security issues—Skywell software [on-line] Skywell Software. Available at: https://skywell.software/blog /top-5-big-data-security-issues/. Accessed 20 July 2020
13. Ekbia H, Mattioli M, Kouper I, Arave G, Ghazinejad A, Bowman T, Suri V, Tsou A, Weingart S, Sugimoto C (2015) Big data, bigger dilemmas: a critical review. J Am Soc Inf Sci 66:1523–1545. https://doi.org/10.1002/asi.23294
14. Mishra S, Tripathy N, Mishra BK, Mahanty C (2019) Analysis of security issues in cloud environment. In: Security designs for the cloud, Iot, and social networking, pp 19–41
15. Mishra S, Mishra BK, Tripathy HK, Dutta A (2020) Analysis of the role and scope of big data analytics with IoT in health care domain. In: Handbook of data science approaches for biomedical engineering, pp 1–23. Academic Press
16. Frizzo-Barker J, Chow-White P, Mozafari M, Ha D (2016) An empirical study of the rise of big data in business scholarship. Int J Inf Manage 36:403–413. https://doi.org/10.1016/j.ijinfomgt.2016.01.006
17. Gholami A, Laure E (2016) Big data security and privacy issues in the CLOUD. Int J Netw Security Appl 8:59–79. https://doi.org/10.5121/ijnsa.2016.8104
18. Dutta A, Misra C, Barik RK, Mishra S (2021) Enhancing mist assisted cloud computing toward secure and scalable architecture for smart healthcare. In: Hura G, Singh A, Siong Hoe L (eds) Advances in communication and computational technology. Lecture Notes in Electrical Engineering, vol 668. Springer, Singapore. https://doi.org/10.1007/978-981-15-5341-7_116
19. Gorodov E, Gubarev V (2013) Analytical review of data visualization methods in application to big data. J Electrical Comput Eng 2013:1–7. https://doi.org/10.1155/2013/969458
20. HP (2014) Internet of Things research study. Retrieved from https://fortifyprotect.com/HP_IoT_Research_Study.pdf. Accessed 15 Apr 2020
21. Jose A, Cartos S (2015) Security and privacy issues of big data. In: Noor J, Mohamed E, Mohamed F (eds) Handbook of trends and future directions in big data and web intelligence. IGI Global, Hershey PA, USA, pp 20–52
22. Jutla D, Bodorik P, Ali S (2013). Engineering privacy for big data apps with the unified modeling language. In: 2013 IEEE international congress on big data, pp 38–45, IEEE. https://doi.org/10.1109/BigData.Congress.2013.15
23. Mishra S, Tadesse Y, Dash A, Jena L, Ranjan P (2019) Thyroid disorder analysis using random forest classifier Intelligent and Cloud Computing. Springer, Singapore, pp 385–390

24. Kindervag J, Balaouras S, Hill B, Mak K (2012) Control and protect sensitive information in the era of big data. Academic Press
25. Leszczynski A (2015) Spatial big data and anxieties of control. Environ Plann D Soc Space 33:965–984. https://doi.org/10.1177/0263775815595814
26. Mahmood T, Afzal U (2020) Big data analytics architecture for security intelligence. In: Proceedings of the 11th international conference on security of information and networks, pp 129–134. https://doi.org/10.1109/NCIA.2013.6725337
27. OWASP (2014) OWASP Internet of Things top ten project. Retrieved August 05, 2014, from https://www.owasp.org/index.php/OWASP_Internet_of_Things_Top_Ten_Project. Accessed 15 Apr 2020
28. Perera C (2013) Context aware computing for the Internet of Things: a survey. IEEE Comm Surveys & Tutorials 16:414–454
29. Popa R, Redfield C (2011) Cryptdb: protecting confidentiality with encrypted query processing. In: Proceedings of the …, pp 85–100. https://doi.org/10.1145/2043556.2043566
30. Strang K, Sun Z (2017) Big data paradigm: what is the status of privacy and security? Ann Data Science (Springer) 4:1–17
31. Sun Z, Strang K, Pambel F (2018) Privacy and security in the big data paradigm. J Comput Information Syst 60:146–155. https://doi.org/10.1080/08874417.2017.1418631
32. Sun Z, Sun L, Strang K (2016) Big data analytics services for enhancing business intelligence. J Comput Information Syst 58:162–169. https://doi.org/10.1080/08874417.2016.1220239
33. Swoyer S (2016) It's official: metadata management is a strategic problem. Transforming data with intelligence [online] Transforming data with intelligence. Available at: https://tdwi.org/articles/2016/11/02/metadata-management-is-a-strategic-problem.aspx. Accessed 11 June 2020
34. Tarekegn G, Munaye Y (2016) Big data: security issues, challenges and future scope. Int J Res Stud Comput Sci Eng. https://doi.org/10.20431/2349-4859.0303001
35. Van Dijk M, Gentry C, Halevi S, Vaikuntanathan V (2010) Fully homomorphic encryption over the integers. In: Advances in cryptology–EUROCRYPT '10, pp 24–43. https://doi.org/10.1007/978-3-642-13190-5_2
36. Venkatraman S, Venkatraman R (2019) Big data security challenges and strategies. AIMS Math 4:860–879. https://doi.org/10.3934/math.2019.3.860

Chapter 3
Big Data Process-Based Security and Privacy Issues and Measures

Vazeerudeen Abdul Hameed, Selvakumar Samuel, and Kesava Pillai Rajadorai

1 Introduction

Privacy and security of data have been of prime concern at all times and the data in different domains needed security. The terms privacy and security are related but not the same. Security ensures that privacy of the data may be protected [1]. A breach of security to the data causes private data to become vulnerable to public access by perpetrators and hence cause different types of losses to the data owner.

Data in different domains has been protected via different security mechanisms. In banking, the usernames and passwords of individuals were stolen by breaching their security and gain monetary advantages [2]. Over time added layers of security such as four digit personal identification number (PIN) via automated telephone banking system helped to control such security breaches. Online chat applications such as Whatsapp uses end-to-end encryption to ensure secure communication between its customers. However, it must be noted that adding security to any domain causes it to slow down the core purpose. Most or all of the security systems consume time. The telecommunication network speed or loss of PIN causes delay in a banking transaction. Encryption and decryption of data over networks cause delay in chat applications like Whatsapp. However, such delays are manageable or acceptable.

V. Abdul Hameed (✉) · S. Samuel · K. Pillai Rajadorai
Asia Pacific University of Technology and Innovation, Kuala Lumpur, Malaysia
e-mail: vazeer@staffemail.apu.edu.my

S. Samuel
e-mail: selvakumar@staffemail.apu.edu.my

K. Pillai Rajadorai
e-mail: kesava@staffemail.apu.edu.my

P. K. Das et al. (eds.), *Privacy and Security Issues in Big Data*, Services and Business Process Reengineering, https://doi.org/10.1007/978-981-16-1007-3_3

2 Challenges in Applying Security Over Big Data

Big data is collected from different data sources such as social platforms like Whatsapp, twitter, and Facebook and machine data such as sensor data from the Internet of things (IoT) and transactional data from business transactions. It must be taken into account that security breach can happen at different points before data is available for processing. Figure 1 shows a simplified presentation of the process behind the collection of big data for analysis. As evident from Fig. 1, security attacks could happen at the data source or the data mapper or the data warehouse or at the data analysis phase. Security mechanisms for data generally depend on encryption and decryption. The very nature of big data as being very big causes the encryption and decryption of data to be extremely slow and hence defeats the purpose of processing big data within a stipulated time.

Efforts have been taken to curb security breach in the areas shown in Fig. 1. The following sections discuss the security issues in each of the areas.

2.1 Data Source

Two major ways to curb security attack at the data source were discussed in [3]. Firstly, cryptographic techniques such as authentication mechanisms and attribute-based credentials were recommended. These methods fundamentally depend upon mathematical encryption and decryption algorithms. Secondly, non-cryptographic methods such as transparency enhancing techniques (TET) and/or intervenability were identified. These methods aim at closer interaction with the end-users to gain

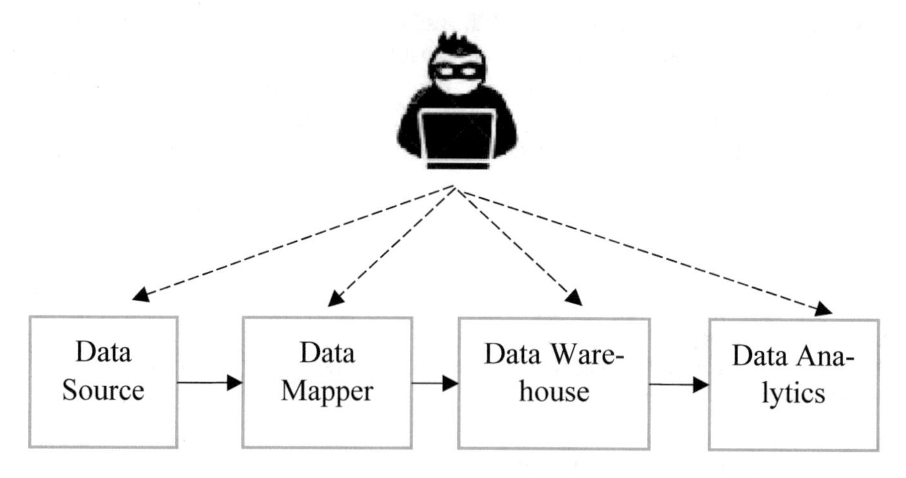

Fig. 1 Big data process

their confidence on disclosing only the necessary data, thereby avoiding the risks of privacy.

2.2 Data Mapper

Research has also been conducted to add security at the data mapper domain. Lightweight encryption techniques were suggested to be introduced between the data source and the data mapper. This layer of protection was named as the secured MapReduce layer [4]. The algorithms clearly depend upon the number of cores used in parallel and the volume of data. The time consumed is of the order of 10^6 ms. However, the research explains the need for security of big data.

2.3 Data Warehouse

Data warehouses pose difficulty to impact security measures due to their inherent size and the lack of any organized structure which is their prime advantage. Security algorithms are mostly avoided in this area.

A key problem in big data environments is the generation of fake data [5]. Fake data could be generated at any of the four areas shown in Fig. 1. A possible infringement into the data warehouse could allow for an uncontrolled generation of fake data which would affect the data analytics phase. Fake data generation could also be promoted for a positive cause. Anonymizing user information in the health sector is an essential step towards safeguarding patients' privacy. This is accomplished via generating fake data [6]. Fake news is a problem which is confronted with the help of several filters. These filters are trained to determine fake news. However, research has been carried out to build models that can generate fake news to train the filters [7]. An algorithm that can predict fake data will be able to prevent a large-scale mishap in the data analysis. The following section presents an algorithm to identify potential fake data.

3 Proposed Algorithm to Eliminate Fake Data

Figure 2 presents a flow of steps involved in the algorithm [8]. Feature extraction and classification of data into true data and fake data have been explained in the following sections.

1. A sufficiently large subset of the data from the warehouse is selected.
2. The data is subjected to conventional methods of data cleansing and preparation and hence made suitable for data mining on the subset. This data is therefore considered as a reliable data subset.

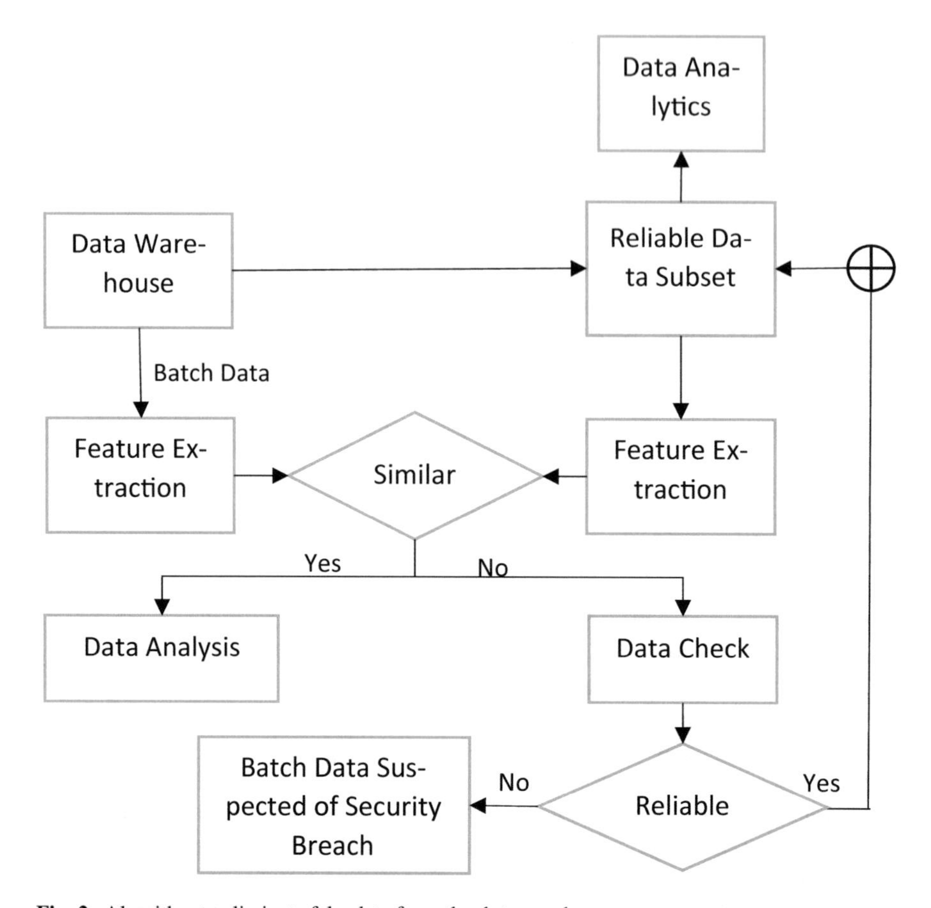

Fig. 2 Algorithm to eliminate fake data from the data warehouse

3. Feature extraction over this data subset is performed and stored.
4. Data from the warehouse is retrieved in batches, and features are extracted. These features are compared with the features recorded in step 3.
5. If the features match, then that batch of data is subjected to conventional data analysis.
6. Should the features show deviation, the batch of data is subjected to further checks or verification.

4 Data Analysis

A given data is analysed to obtain several statistics. The statistics could range from simple measures such as a mean, median or mode to in-depth analysis such as chi-square statistical measure [9]. However, any statistical measure would be wrong if the given data was corrupted with false data. The following section suggests a feature extraction method that can be used by the algorithm explained in Fig. 2 to eliminate false data.

5 Feature Extraction

The concept of feature extraction has been widely discussed in several researches. A wide number of mathematical models for feature extraction were developed by researchers, which when combined with the enormous computational power of the latest technologies in computing have yielded good results in feature extraction over big data [10]. The following section discusses some of the possible methods of feature extraction.

5.1 *Feature Extraction with Moment Functions*

Feature extraction is often accomplished via some mathematical models. Moment functions are such mathematical models that have been widely used for feature extraction. Moments were first contributed by David Hilbert [11]. Two major classifications of moments, namely orthogonal and Cartesian moment functions, have been contributed via various researches, and these functions have been used to solve a large number of problems. The orthogonal moment invariants have been largely used in applications that require feature extraction from the given data and reconstruction of the original data using the features. The orthogonal moment invariants are also known for extracting unique features in the given data. Unlike orthogonal moment invariants, the Cartesian moment functions are computationally very simple as they are always real valued. However, Cartesian moment functions have less or no capability of reconstruction of data. In this research, Hu moment invariants [12], which are Cartesian in nature, have been applied. A Hu moment function of order p is formulated as in Eq. (1). The given function is of one dimension. It could be expanded with cascaded summations over higher dimensions. Hu moment-invariant functions were originally defined as seven functions as listed in Eq. (2). These seven functions were further extended via algorithms stated in [13]. The n_{pq} is two-dimensional moments which are obtained by normalizing M_{pq}. These invariants were originally used for feature extraction in images and hence were defined in two dimensions.

$$M_p = \sum_p x^p I(x) \tag{1}$$

$$I_1 = n_{20} + n_{02}$$
$$I_2 = (n_{20} - n_{02})^2 + 4n_{11}^2$$

$$
\begin{aligned}
I_3 &= (n_{30} - 3n_{12})^2 + (3n_{21} - n_{03})^2 \\
I_4 &= (n_{30} + n_{12})^2 + (n_{21} + n_{03})^2 \\
I_5 &= (n_{30} - 3n_{12})(n_{30} + n_{12})[(n_{30} + n_{12})^2 - 3(n_{21} + n_{03})^2] \\
&\quad (3n_{21} - n_{03})(n_{21} + n_{03})[3(n_{30} + n_{12})^2 - (n_{21} + n_{03})^2] \\
I_6 &= (n_{20} - n_{02})[(n_{30} + n_{12})^2 - (n_{21} + n_{03})^2] + 4n_{11}(n_{30} + n_{12})(n_{21} + n_{03}) \\
I_7 &= (3n_{21} - n_{03})(n_{30} + n_{12})[(n_{30} + n_{12})^2 - 3(n_{21} + n_{03})^2] \\
&\quad - (n_{30} - 3n_{12})(n_{21} + n_{03})[3(n_{30} + n_{12})^2 - (n_{21} + n_{03})^2]
\end{aligned} \tag{2}
$$

In order to illustrate the use of Hu moment-invariant functions, a random sample data was created. The data involved 350 samples of two attributes, namely the age and height of a tree breed in years and feet, respectively. Figure 3 shows the plot of selected 300 samples which are considered to be the reliable data subset as explained in Fig. 2. The Hu moment invariants for these 300 data points were estimated and recorded in Table 1. Figure 4 shows the plot of the 300 samples in blue plus the remaining 50 reliable samples in red. The corresponding Hu moment invariants were recorded in Table 1. It can be observed that the deviation in the values of the moment invariants is very small. However, Fig. 5 shows the plot of 300 samples of the reliable data subset plus 50 samples of false data. From Table 1, it is evident that the Hu moment invariants show larger deviation for the false data. It must be noted that not all moment invariants, particularly that of the lower-order functions may not show a wide deviation but the higher-order moment invariants are good at capturing the deviations in the data. The threshold of acceptable deviation depends on the data analyst.

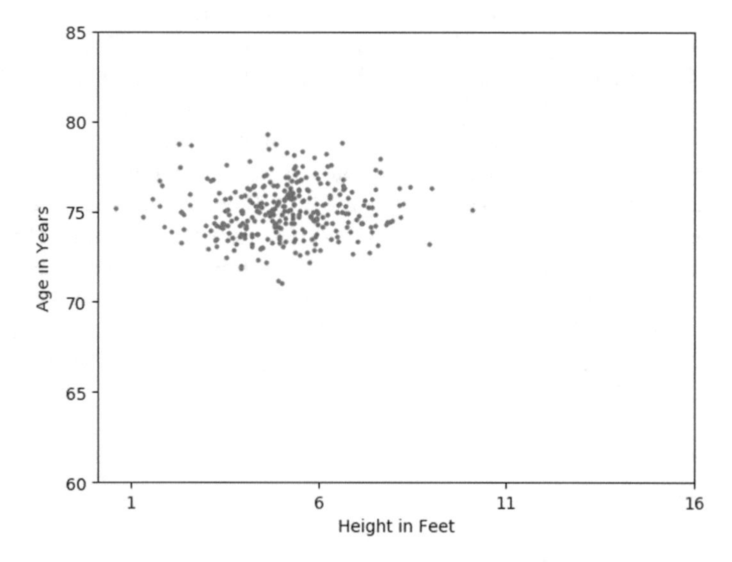

Fig. 3 Reliable data

Table 1 Measure of moment invariants for three data plots

Moment invariant	Data in Fig. 3	Data in Fig. 4	Data in Fig. 5
I1	3.1581809	3.15664111	3.15547215
I2	7.41655505	7.41157539	6.41291343
I3	14.3257418	14.33247253	15.1527896
I4	13.51787666	13.5715346	13.29225576
I5	27.56351383	27.69296357	28.0478668
I6	17.32312685	17.45619576	16.03168856
I7	−27.62063344	−27.65665848	−27.43965211

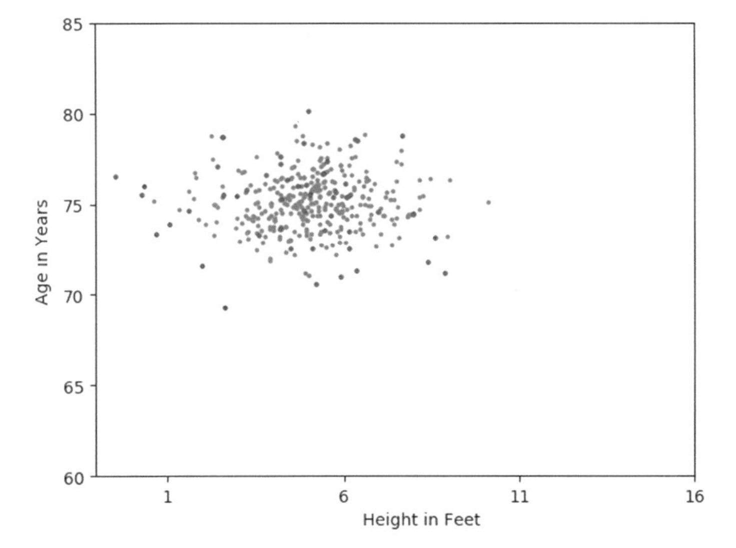

Fig. 4 Reliable data with true data

An algorithm with Hu moment-invariant-based feature extraction of the data was used to detect the possible presence of fake data in a pool of data. The approach yielded abilities to differentiate true and fake data in a data set. It is vital to note that the performance of the algorithm can vary with the nature of the data. It is recommended that other feature extraction methods such as Zernike moments, Legendre moments or the like could be applied over the data to study the possibilities of better feature extraction in the attempt towards differentiating true and fake data. The following section presents a different approach to classify true and fake data.

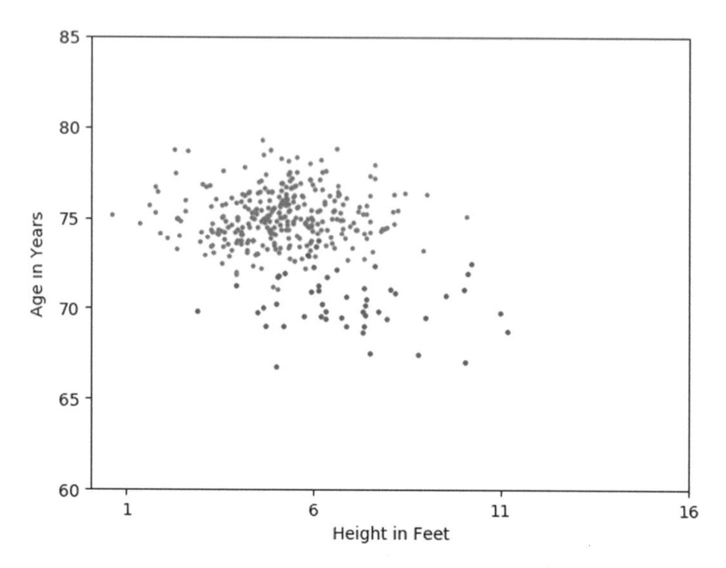

Fig. 5 Reliable data with false data

6 Feature Extraction with Neural Networks

The concept of neural networks was first explained by Frank Rosenblatt in 1958. Neural networks are an attempt to simulate the structure of the neurons in a human brain that helps to learn patterns based on experience. For instance, a human being encounters several experiences in life, since birth which enables him to make the right decisions. Hence, a human brain learns, updates and makes decisions. The three keywords, namely learn, update and decide, are vital for an effective neural network.

The first neural network was called the perceptron which replicates a single neuron in the human brain. A perceptron is a single neuron that accepts a set of inputs and produces an output based on its past learning experience. The structure of a perceptron is shown in Fig. 6. In the figure, the set of inputs f1, f2, … fn are the set of input features to the neuron. The features are multiplied by a set of random weights w1, w2, … wn. This can be understood as determining the importance of a specific feature. The higher the weight the more important will be the feature to determine the result and vice versa. The aggregate of the features along with a bias is subjected to an activation function. The result of the activation function could be as simple as a true (numerically equal to one) or false (numerically equal to zero) which is the decision made by the perceptron.

A perceptron is subjected to a training phase and validation phase before it can be reliably used. A module named error and weight update is used while training a perceptron. This module helps a perceptron to learn and update the weights for the feature vector. Every new perceptron starts with a random set of weights for the feature vector. The perceptron is trained to recognize the pattern of the feature vector

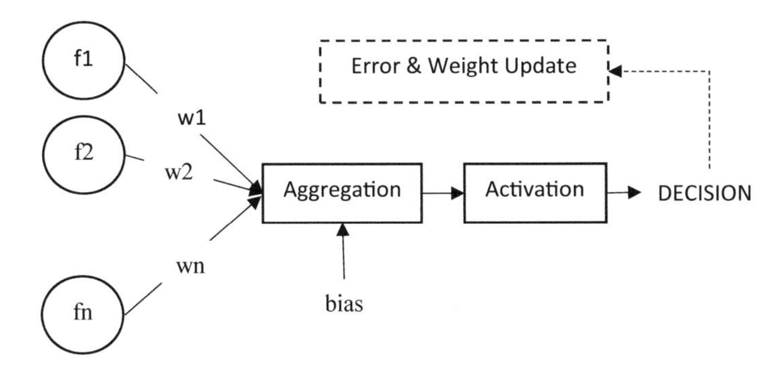

Fig. 6 A perceptron in a neuronetwork

by forcefully updating the weights for a given set of features and their corresponding decision. Once the perceptron is sufficiently trained, the error and weight update module can be disabled and the perceptron is put into use for decision-making.

Practical applications could deploy a large number of perceptron, interconnected in different shapes to make important decisions. Such a network of perceptron is called as a neural network. Several architectures of neural networks have been proposed. Figure 7 through Fig. 10 presents some of the commonly used neural network architectures. Every node (circle) in the network represents a perceptron. The most commonly used architecture is the feed forward neural network.

The nodes in Fig. 7 through Fig. 10 have been colour coded as white, light grey and dark grey. The white perceptron are inputs, the light-grey perceptron are data processing nodes/hidden nodes and the dark grey nodes are the results of the feature extraction. A simple feed forward network in Fig. 7 has one input layer, one hidden layer and one output layer. The number of nodes in the input layer decides the number of input features given to the network. Similarly, the number of nodes in the output layer decides the number of characteristics understood by the network based on the inputs. Changes to this simple feed forward network yields a deep learning network shown in Fig. 7, where two hidden layers are found. Adding several hidden

Fig. 7 Feed forward network

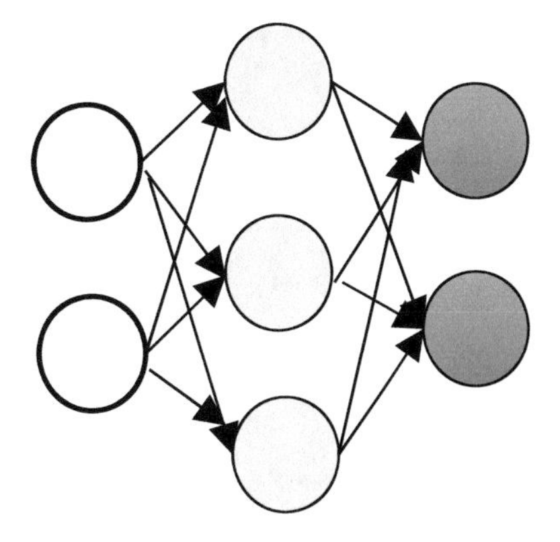

layers improves the learning capacity of the neural network, thereby producing better results. However, the network may become slow due to increased processing by the large number of perceptron. Hence, deep learning neural network in Fig. 8 is more accurate but less efficient when compared to simple feed forward network [14]. Figure 9 shows a recurrent neural network which is very similar to the deep learning neural network in Fig. 8 but with an added memory component to each of the hidden layer nodes [15]. Neural networks in general do not remember past data that was processed by them unlike human beings. A recurrent neural network simulates this memory factor by storing the previous results of a perceptron into itself for future use. This is symbolically shown by the self-directing arrows in the nodes in Fig. 9. Figure 10 presents a binary decision-making neural network commonly known as state vector machines. Hence, the decision of this neural network is limited to a True/False or zero/one type of results. Hence, several simple/complex architectures of neural networks have been made possible.

The neural networks could be trained to extract features in a data. This knowledge can be used to differentiate true and fake data. The knowledge is stored within the

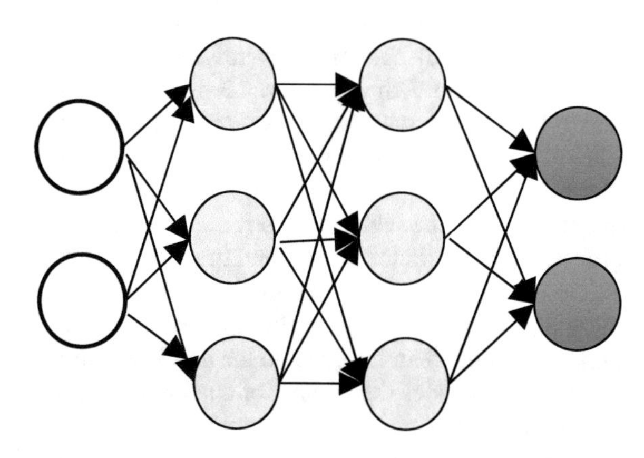

Fig. 8 Deep learning neural network

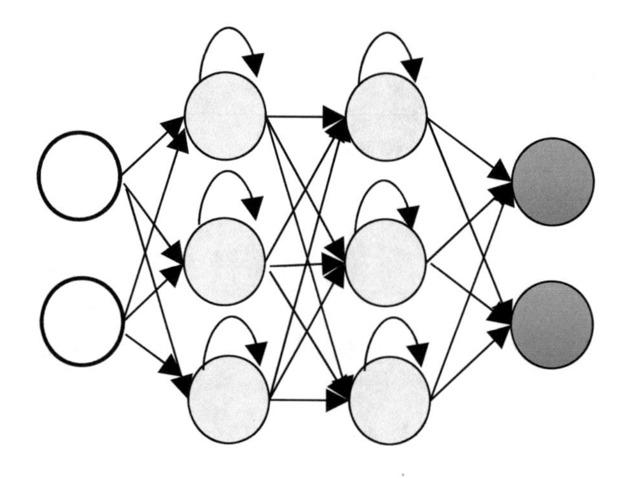

Fig. 9 Recuurent neural network (RNN)

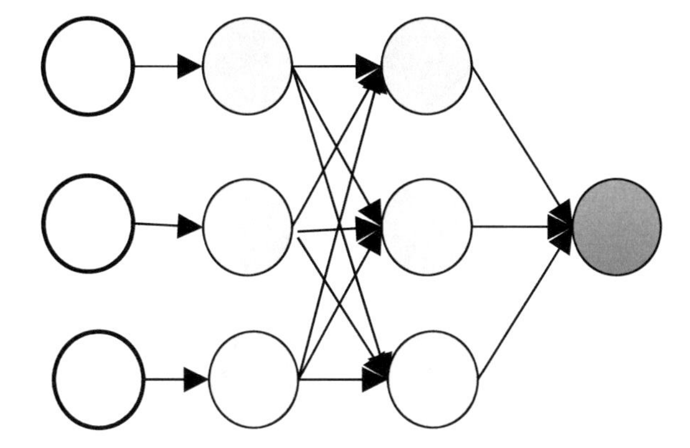

Fig. 10 State vector machine (SVM)

neurons of the network and do not require any human intervention once the network has been deployed. The following section illustrates the use of a neural network to understand how true and fake data could be classified with the help of a neural network.

A data sample of 10,000 records was used to illustrate the use of neural networks for feature extraction over big data. The data involves two parameters, namely height and weight of a population of people. The data has been classified into true data and false data for each pair of height and weight. The following feed forward network in Fig. 11 was implemented. As explained above, the neural network was trained using a portion of the sample data over several iterations called epochs. Each epoch involved training of the network with a large batch of data. This way, the network was trained to learn the characteristics of the true and the false data.

The training was completed after 2122 epochs. The number of epochs would vary with the number of layers of the hidden layer and the number of nodes within each hidden layer. The training will involve a random selection of data due to which the network may require different number of epochs when it is subjected to training each time. During the training, a loss function explains how well the network is correcting itself through the learning process. This can be observed from the loss

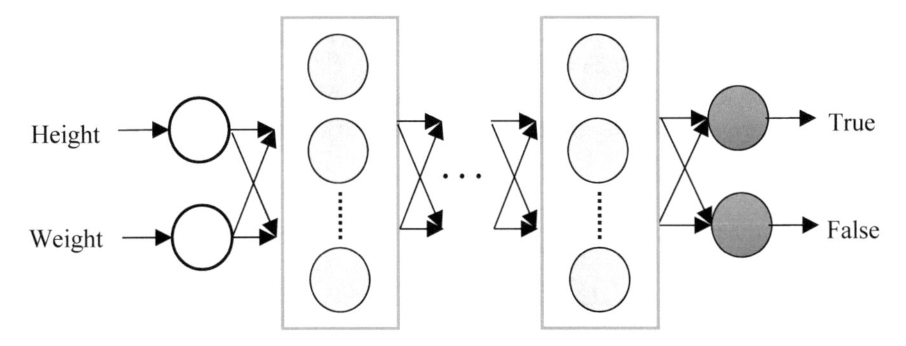

Fig. 11 Feature classifier

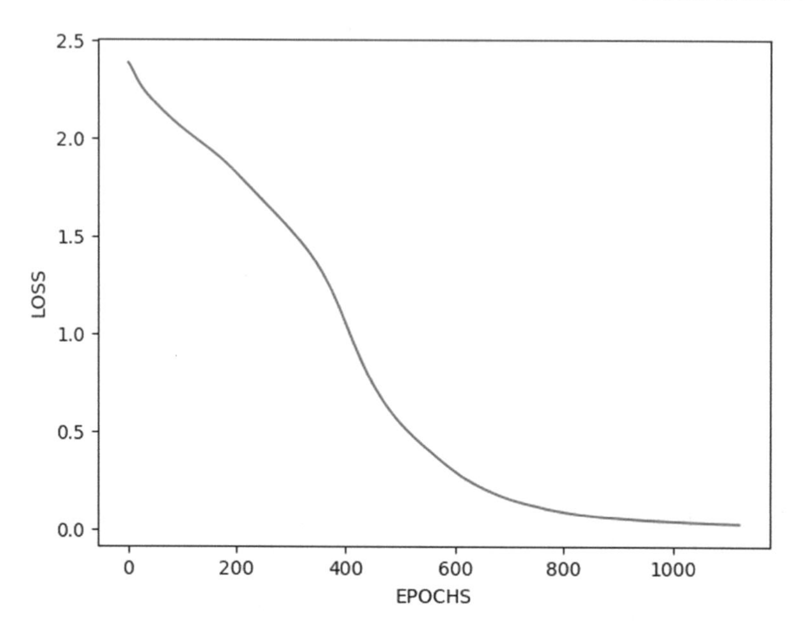

Fig. 12 Loss function

function plotted as a graph. The loss function for the training of the neural network in Fig. 11 is shown in Fig. 12. The graph shows that the loss function was a value well above 2.0 at the beginning of the training. The training ended when the loss reduced to 0.012. Ideally the loss can be zero, but in real time, every neural network is prone to mistakes. A neural network that is perfectly trained is far from ideal.

Once training was completed, the network was subjected to a random test with some data that were not introduced to the network during its training. A batch of 10,000 records was input to the neural network. The data had a mix of valid and invalid pair of height to weight. The following confusion matrix in Table 2 was produced. The table explains the true positive, true negative, false positive and the false negative. 99.13% of the data has been correctly classified as valid and invalid pair by the neural network, while 0.87% of the data has been wrongly classified. The tolerable threshold of error could be a deciding factor to determine if further training of the neural network may be needed.

Neural networks have long been used to solve several optimization problems. An artificial neural network was trained to differentiate between potential fake data

Table 2 Confusion matrix

Actual	Predicted	
	Valid pair	Invalid pair
Valid pair	5768	52
Invalid pair	35	4145

and true data. After sufficient training and testing, the neural network was found to demonstrate capabilities of differentiating true and fake data. A feed forward neural network was put to use in the experiment in this section. However, it is recommended that there are several other neural network architectures such as convolutional neural networks (CNN), recurrent neural network (RNN) and others which could be studied for their effectiveness in better feature extraction towards identifying fake data. The following section presents a different approach towards handling true and fake data.

7 Feature Extraction with Clustering

Clustering algorithms are a vital set of machine learning solutions that have been used to solve a large number of optimization problems. Clustering of data has been accomplished via several algorithms such as K-means clustering, mean shift clustering [16], density-based spatial clustering of applications with noise (DBSCAN) [17], expectation maximization clustering [18], agglomerative hierarchical clustering [19] and many more. K-means clustering is a commonly known technique that has been deployed to solve several problems in machine learning. This algorithm could be deployed to extract features in a data set. These features can be used to detect potential fake data from the true data in a data set [20].

Figure 13 illustrates the steps involved in classifying potential fake data from true data. The process involves two stages. Stage—1 is a preparatory stage. A randomly selected set of true and fake data points are grouped into clusters. There could be several true data clusters and fake data clusters to begin with. Stage—2 is an iterative phase. In this phase, every new data point is fitted into one of the existing clusters. However, some data points may be weakly connected to all of the existing clusters. In such situations, a new cluster is created. Hence, the model evolves with time to accommodate every new data point to one of the existing clusters. The number of clusters will become optimized automatically when there are sufficient number of clusters such that any given data point becomes strongly connected to one of the existing cluster centroids [21, 22]. Now the clusters act as deciding factors to differentiate a true data point from a fake data point. Like every other problem-solving techniques that involve optimization, the approach helps to determine an optimized solution to classifying fake data from the true data. Hence, the performance of this approach depends largely on the initial stages of training where the system learns to differentiate a true cluster centroid from a fake centre centroid.

The approach was modelled into a prototype to visualize the working of the approach. A sample three-dimensional data of records were used to demonstrate the working of the approach to classify the true and fake data using clustering techniques. Figure 14 shows an initial plot of manually selected features classified into clusters. The blue cluster of points was set to be the true data points, and the red and green clusters were set to be expected fake data points [23]. The graph presents only few data points for clear visualization. Hence, Fig. 14 corresponds to the Stage—1 in Fig. 13.

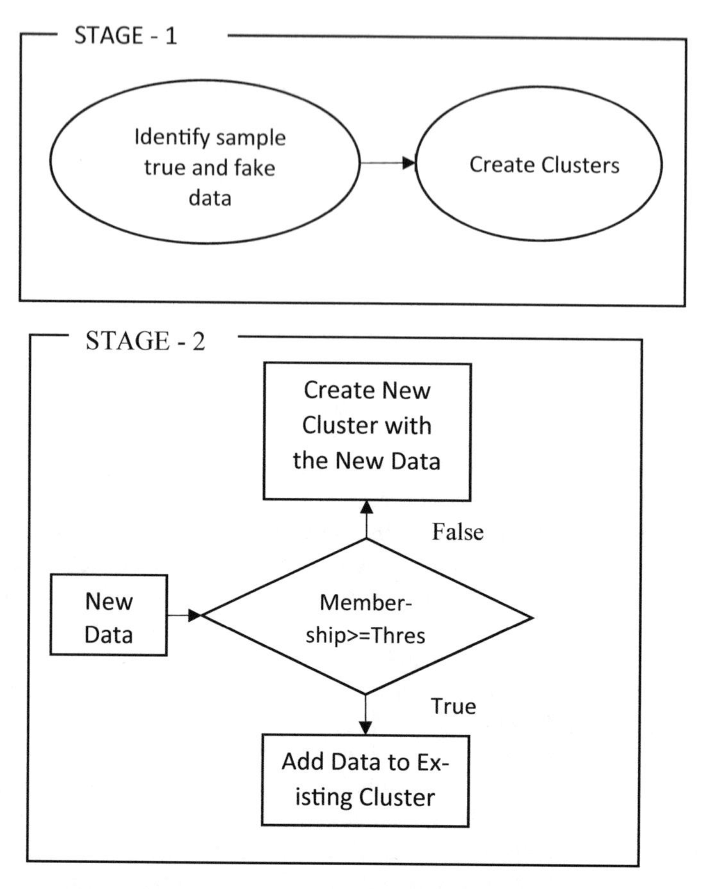

Fig. 13 Classification of fake and true data with K-means clustering

Figure 15 shows the plot of additional eighty data points from a pool of data to be classified into fake data and true data. It can be seen that a new cluster has emerged between the blue and the green regions of the original plot in Stage—1, Fig. 14. The modelled prototype converged into fitting a set of five hundred data points into one of the four clusters. It was verified that approximately 70% of the true data were correctly clustered into the true data group, which explains that 30% of the true data were lost in the clusters of the fake data. Similarly, over 80% of the fake data were grouped into one of the fake data clusters while 20% of the fake data were absorbed by the true data cluster.

K-means clustering algorithm was used to model a solution to differentiate between true and fake data. An experiment to classify true and fake data points was presented. The data points were composed of three features, and hence, they were visualized as three dimensional graphs. It is recommended that further study could be made on the impact of increasing the features used to represent the data

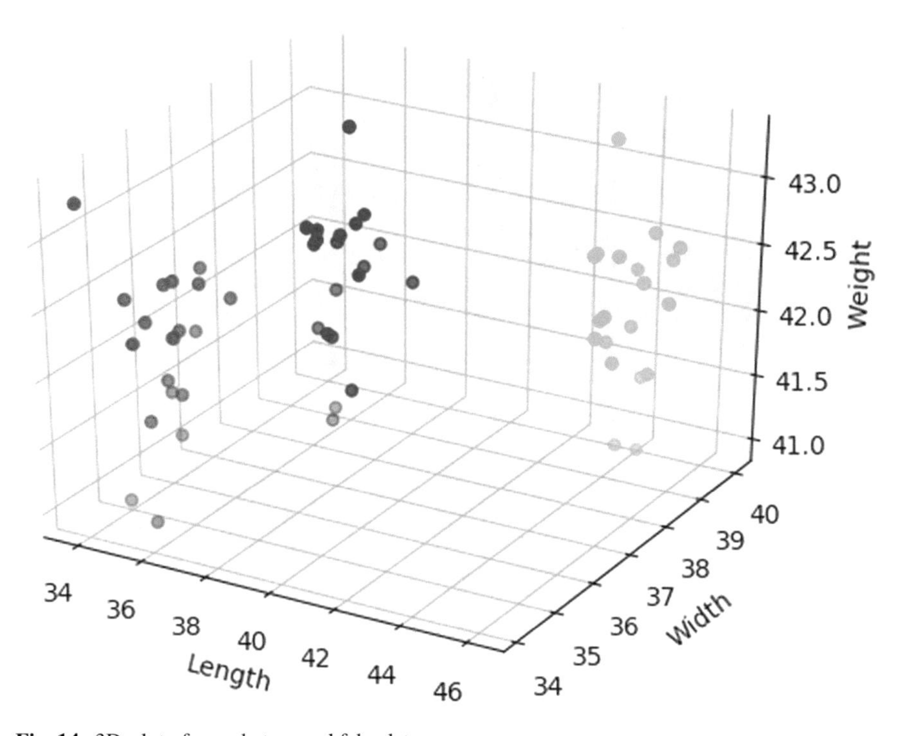

Fig. 14 3D-plot of sample true and fake data

points. There are a large number of different clustering algorithms that could be applied in this approach to classify true and fake data [24].

8 Conclusion

The chapter presents an understanding of the various approaches taken by researchers to address security of big data. Fake data generation is considered to be one of the major forms of attacking big data. True data can be separated from fake data by deploying machine learning algorithms to extract features in the data. Features of true data are certainly different from the features of fake data. Hence, a viable feature extraction method could be identified for this purpose. This chapter discusses three plausible ways in which fake data could be separated from true data with the help of machine learning algorithms to extract the features in the data. The suggested methods could be varied by deploying alternative machine learning algorithms to obtain better results.

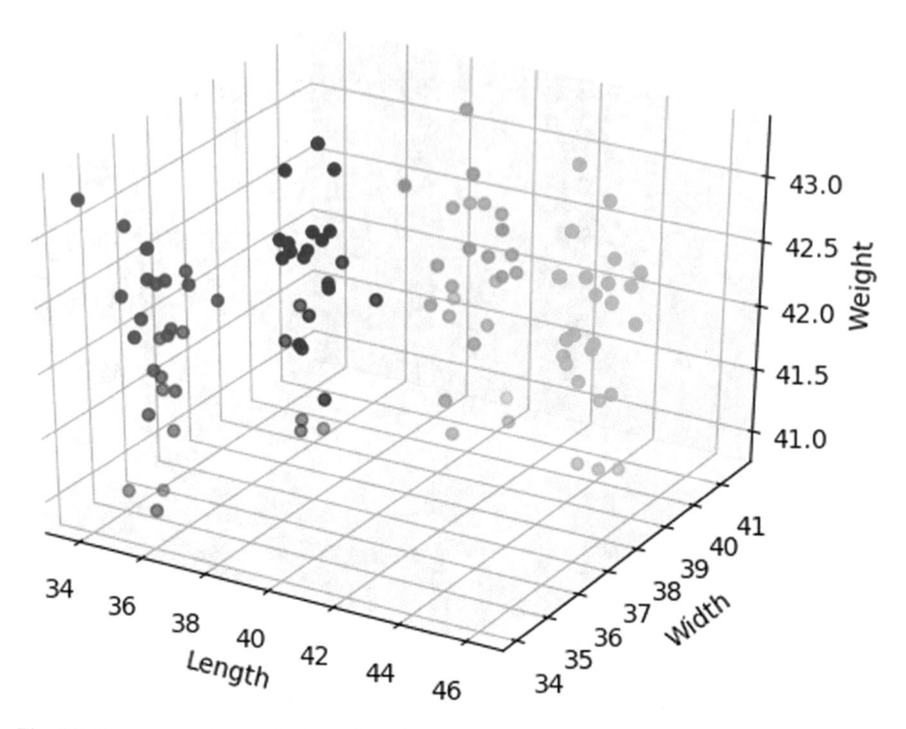

Fig. 15 Emergence of new cluster of fake data

In summary, although several measures could be taken to curb the impact of fake data in big data processing systems, the choice of the measure is dependent upon the nature of the data being processed. Hence, careful study on the appropriateness of a specific technique is essential before deploying the technique to handle the fake data in big data processing environments.

References

1. Mishra S, Sahoo S, Mishra BK (2019) Addressing security issues and standards in Internet of things. In Emerging trends and applications in cognitive computing. IGI Global, pp 224–257
2. Rath M, Mishra S (2019) Advanced-level security in network and real-time applications using machine learning approaches. In: Machine learning and cognitive science applications in cyber security. IGI Global, pp 84–104
3. Salas J, Domingo-Ferrer J (2018) Some basics on privacy techniques, anonymization and their big data challenges. Math Comput Sci 12:263–274. https://doi.org/10.1007/s11786-018-0344-6
4. Jain P, Gyanchandani M, Khare N (2019) Enhanced secured map reduce layer for big data privacy and security. J Big Data. https://doi.org/10.1186/s40537-019-0193-4
5. Sivarajah U, Kamal M, Irani Z, Weerakkody V (2017) Critical analysis of big data challenges and analytical methods. J Bus Res 70:263–286. https://doi.org/10.1016/j.jbusres.2016.08.001

6. Piacentino E, Angulo C (2020) Generating fake data using GANs for anonymizing healthcare data. Bioinformatics and Biomed Eng 406–417. https://doi.org/10.1007/978-3-030-45385-5_36
7. Dandekar A, Zen R, Bressan S (2017) Generating fake but realistic headlines using deep neural networks. Lect Notes Comput Sci 427–440. https://doi.org/10.1007/978-3-319-64471-4_34
8. Mishra S, Tripathy HK, Mallick PK, Bhoi AK, Barsocchi P (2020) EAGA-MLP—an enhanced and adaptive hybrid classification model for diabetes diagnosis. Sensors 20(14):4036
9. Mishra S, Mahanty C, Dash S, Mishra BK (2019) Implementation of BFS-NB hybrid model in intrusion detection system. In: Recent developments in machine learning and data analytics. Springer, Singapore, pp 167–175
10. Mohapatra R, Mishra S, Mohapatra T (2012) Coverage problem in wireless sensor networks. Comparat Cytogenet 2(1):67–72
11. Hilbert D (1993) Theory of algebraic invariants. Cambridge Univ. Press, Cambridge
12. Ming-Kuei Hu (1962) Visual pattern recognition by moment invariants. IEEE Trans Inf Theo 8:179–187. https://doi.org/10.1109/tit.1962.1057692
13. Abdul Hameed V (2016) Determination of the appropriate geometric moment invariant functions for object recognition. Ind J Sci Technol. https://doi.org/10.17485/ijst/2016/v9i21/95209
14. Georgevici A, Terblanche M (2019) Neural networks and deep learning: a brief introduction. Intensive Care Med 45:712–714. https://doi.org/10.1007/s00134-019-05537-w
15. Ceni A, Ashwin P, Livi L (2019) Interpreting recurrent neural networks behaviour via excitable network attractors. Cognit Comput 12:330–356. https://doi.org/10.1007/s12559-019-09634-2
16. Ren Y, Kamath U, Domeniconi C, Zhang G (2014) Boosted mean shift clustering. machine learning and knowledge discovery in databases pp 646–661. https://doi.org/10.1007/978-3-662-44851-9_41
17. Duan H, Wang X, Bai Y et al (2019) Integrated approach to density-based spatial clustering of applications with noise and dynamic time warping for breakout prediction in slab continuous casting. Metall Mater Trans B 50:2343–2353. https://doi.org/10.1007/s11663-019-01633-w
18. Jin X, Han J (2017) Expectation maximization clustering. encyclopedia of machine learning and data mining pp 480–482. https://doi.org/10.1007/978-1-4899-7687-1_344
19. Roux M (2018) A comparative study of divisive and agglomerative hierarchical clustering algorithms. J Classif 35:345–366. https://doi.org/10.1007/s00357-018-9259-9
20. Mallick PK, Mishra S, Chae GS (2020) Digital media news categorization using Bernoulli document model for web content convergence. Pers Ubiquit Comput. https://doi.org/10.1007/s00779-020-01461-9
21. Mishra S, Mallick PK, Jena L, Chae GS (2020) Optimization of skewed data using sampling-based preprocessing approach. Front Publ Health 8:274. https://doi.org/10.3389/fpubh.2020.00274
22. Sahoo S, Mishra S, Panda B, Jena N (2016) Building a new model for feature optimization in agricultural sectors. In: 2016 3rd international conference on computing for sustainable global development (INDIACom), New Delhi, pp 2337–2341
23. Hameed VA, Shamsuddin SM (2013) Generalized rotational moment invariants for robust object recognition. Int J Adv Soft Compu Appl 5(3)
24. Mishra S, Mallick PK, Tripathy HK, Bhoi AK, González-Briones A (2020) Performance evaluation of a proposed machine learning model for chronic disease datasets using an integrated attribute evaluator and an improved decision tree classifier. Appl Sci 10(22):8137

Chapter 4
Exploring and Presenting Security Measures in Big Data Paradigm

Astik Kumar Pradhan, Jitendra Kumar Rout, and Niranjan Kumar Ray

1 Introduction

In the past few years, the word 'Big Data' has achieved popularity in the competitive business world [1]. Many organizations like e-commerce, banks, retail, and hospitals generate a huge amount of data every day. Not only humans but machines also contribute to the process of data generation through Website logs, TV streaming, etc., and in addition to this excessive use of social media and smart phone's also produces a massive amount of data every minute and every day. In this competitive world, the big reputed companies depend on data management, integration, and data abstraction for effective decision making which actually required in the growth and development of the companies. Therefore, Big Data analytic is very essential for the success of the company in recent times. Big e-commerce companies like eBay, Flipkart, and Amazon use the recommendation systems which is one of the applications of Big Data analytic, to indicate the goods to various customers based on their purchasing patterns, and this leads to inference attacks. Although big data analytic is helpful in decision making, it also leads to serious privacy and security issues [2]. Before studying the privacy and security in Big Data let us understand first 'What is Big Data?'.

1.1 What Is Big Data?

The data may be numbers, characters, or symbols in which computer operations are performed in order to process and transmitted in the form of electrical signals and recorded in optical, mechanical, or magnetic media recorder. Big Data is data as well, but with an immense scale and dimension. Big Data is a notion used to describe a

A. K. Pradhan (✉) · J. K. Rout · N. K. Ray
School of Computer Science and Engineering, KIIT University, Bhubaneswar, Odisha, India

P. K. Das et al. (eds.), *Privacy and Security Issues in Big Data*, Services and Business Process Reengineering, https://doi.org/10.1007/978-981-16-1007-3_4

dataset that, in terms of volume, is enormous and can expand exponentially over time. Simply put, Big Data is more complex, and large datasets that cannot be stored and processed efficiently by any of the traditional database management systems [3–5].

1.2 Examples of Big Data

To grasp the definition of Big Data more precisely, here are some of the examples.

1. ***Marketing***: Amazon ads based on what you buy. Actually, Amazon got sucked into the advertising business by the sheer amount of customer data at its disposal. This company has accumulated a large number of information related to the buying habits of millions of people, where those purchases are delivered, and what kind of payment cards they use. Amazon allows different marketing companies to access its self-service ads portal so that they can purchase different ad campaigns.

2. ***Social Media***: More than five hundred terabytes of new data are entered into social media databases such as Facebook, Instagram every day. These new data are created in the form of photographs, uploaded videos, comments, etc.

3. ***Transportation***: Maps to applications. The majority of smart phone users depend on their devices for directions, i.e., to know about the traffic road patterns. By analyzing the big amount of traffic data, we can find out the road which is busy and takes more time to reach the landing place or terminus.

1.3 Different Forms of Big Data

Now that we are on board with what Big Data is, let us take a look at various types of Big Data [6]. Big Data can primarily be divided into three forms:

1. Structured
2. Uunstructured
3. Semi-structured

1. Structured

The data that can be stored, processed, and retrieved in a fixed format is referred to as structured data (as shown in Table 1). In reality, it refers to well-organized information that can be quickly and easily accessed via a simple algorithm search engine.

Structured data Example: - A good example of structured data is the employee table in a company database as the employee information such as employee ID, gender, and department are present in an ordered manner.

Table 1 Example of Structured data (employee Database)

Employee_id	Employee name	Gender	Salary	Department
1,950,002	Ranjan	Male	50,000	CSE
1,950,003	Anupama	Female	55,000	IT
1,950,004	Ram	Male	60,000	IT
1,950,004	Astik	Male	65,000	IT

2. Unstructured

Any data that does not support any particular type of structure is known as unstructured data. Unstructured data is time-consuming in nature as it required a huge amount of time to process and analyze the data, and this is very difficult also as the type of the data is unknown.

Unstructured data Example: - The output return by the Google search engine (as shown in Fig. 1) is a very good example of unstructured data because here the type and structure of the data are unknown.

3. Semi-structured

Semi-structured is the third type of Big Data. This form of data includes both the forms of data that are listed above. Put simply, the information that supports both organized and unstructured forms is called semi-structured data. In reality, semi-structured data is structured but not specified. Data in an XML file is a clear example of semi-structured data. The data has not been listed in a specific database here, but

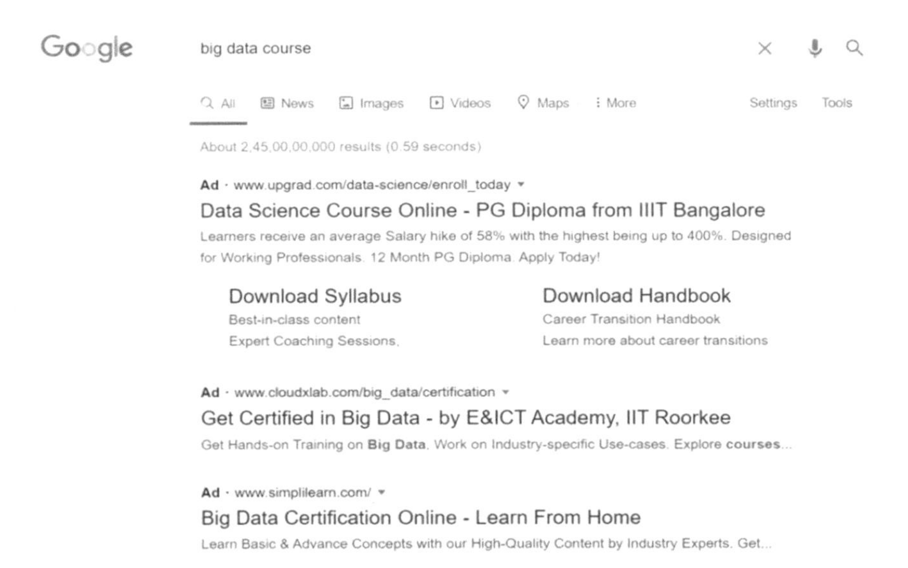

Fig. 1 Image of google search engine

Personal data stored in a XML file.

<rec><name>Raj Kumar</name><sex>Male</sex><age>40</age></rec>

<rec><name>Mohan Gope</name><sex>Male</sex><age>30</age></rec>

<rec><name>Jitu Pradhan</name><sex>Male</sex><age>18</age></rec>

Fig. 2 Example of semi-structured data

still contains very vital information [7]. A snapshot of the semi-structured data is shown in Fig. 2.

1.4 Characteristics of Big Data

We have come to the limit of data forms. Let us explore the features of Big Data. In recent years, Big Data has been characterized by the three Vs but now there are five Vs available, which are often referred to as Big Data characteristics [8].

1. Volume

Volume is a huge amount of data and 'Big Data' is a name which itself associated with the enormous size. The size of the data is needed to determine the significance of the data. Data volume is required to determine whether a particular data is Big Data or not. Thus, the volume is one of the main features of Big Data that supply information pertaining to the scale of the data [9].

2. Variety

Variety relates to structured, unstructured, and semi-structured data gathered from various sources, such as social media, business processes, and networks. It also refers to the heterogeneous sources. The capability to classify incoming data into different categories is called variety [10].

3. Velocity

Velocity is the estimate of the speed at which the data enters. Velocity describes the rate at which the data is captured, shared, and generated. The data with high velocity required distributed processing techniques. Facebook posts, Twitter messages are good examples of data that is generated with very high velocity.

4. Veracity

Veracity is the uncertainty and inconsistency in the data. It refers to data that is available but sometimes gets messy due to which accuracy and quality both are very difficult to maintain. Veracity is the quality of data that is being analyzed. The data with high veracity are valuable because they help to generate meaningful (overall) results [11]. However, the data with low veracity contains a high percentage of meaningless data.

5. Value

The hidden information can be extracted from the data which is valuable and the bulk of data without having any value is not useful for any company. Value is the most significant aspect of Big Data since it explains whether or not the data available is useful.

So far we have addressed that Big Data is an area that offers numerous ways of analyzing and extracting information and secret patterns. It also helps to cope with the dynamic and larger sizes of data sets. Data has greater predictive strength in many situations, while data with greater complexity contributes to higher false discovery rates [12]. But currently due to the key concepts like volume, variety, velocity, value, and veracity which are associated with Big Data, privacy and security are very important as well as biggest challenges in this field. The security model of the Big Data is not recommended for complex applications and as a result of which by default it gets disabled. But in the absence of this model, the data can be easily compromised. So let us understand various issues related to security and privacy in Big Data.

2 Privacy in Big Data

In Big Data, privacy is the right to monitor the collection, organization, and use of personal information. The right of a group or person to prevent other individuals from knowing information about themselves than those they give the information to is known as the privacy of information [13]. Different government and private business companies are continuously generating and collecting huge amounts of data, and because of the enormous size of the data, it is tough to maintain the data privacy of different customers. Besides this, some of the organizations collect the personal information of the users in order to add value to their business. It is very important that the applications of distinct platforms should be able to provide the safety and privacy of user data.

2.1 Privacy Threats in Big Data

If personal data is present in the public domain, since the data is owned by the data holder, it may lead to a challenge to an individual's privacy. Here, a smartphone application, website, social networking application, etc., maybe the data holder [14]. The data holder is responsible for maintaining the integrity of the data of multiple users. In addition to this, multiple users knowingly or unknowingly contribute to data leakage. Some of the threats to privacy in Big Data include (1) discrimination; (2)

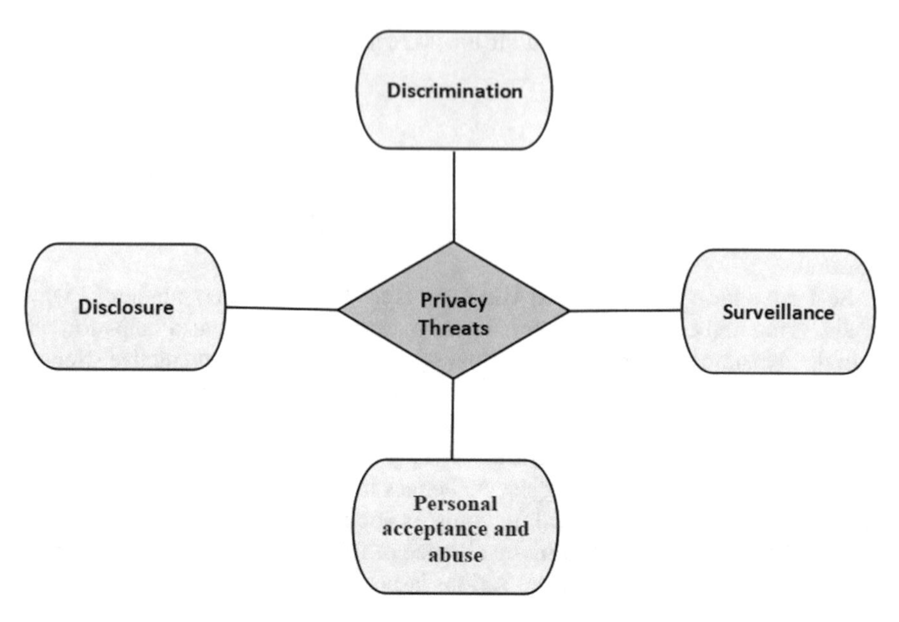

Fig. 3 Privacy threats in big data

disclosure; (3) personal acceptance and abuse; and (4) surveillance, which has been shown in Fig. 3.

1. Discrimination

Discrimination is the inequality or bias which generally happens when some personal information of an individual is disclosed. Different companies for their own benefit share the personal data of different users to other companies without providing the information to the users [15].

2. Disclosure

Disclosure is an action of making new or secret information known. Consider a hospital that contains personal information related to the patients like the name, address, pin, phone number, and disease. If the hospital which is a data holder, in this case, share the personal data of the patients to the third party for analysis purpose of the disease. The third-party's data analyst can map this data with external data sources that are publicly accessible, such as census data, and can classify individuals suffering from any illness. So here the personal data is exposed by the data owners.

3. Personal Embracement and Abuse

The leakage of an individual's personal details can lead to personal acceptance and even violence. A cybercriminal, for example, might spoof a legitimate or approved company email address very easily and request confidential information to be sent to them. The consumer would submit the information that may contain financial details or sensitive information on pricing without being aware of it [16].

4. Surveillance

Surveillance is the close monitoring of private data in order to recognize a person's emotions. Marketing today is about trying to influence what individuals think about a commodity. Several business firms, including e-commerce, retail, etc., research their customers shopping patterns and try to come up with different deals and value-added services. Different social media platforms offer suggestions for new friends, people to follow, places to visit, etc., by using sentiment analysis and statistical analysis of the personal data of the number of users. This is only possible when companies track their customer's transactions on an ongoing basis. This is a significant threat to privacy, as no person embraces surveillance.

2.2 *Privacy Issues in Big Data*

Big Data analytic is a key to understand the behavior of the customers and to make smart decisions for the growth and development of the company. Privacy issues in Big Data are only possible if the organization is unable to manage the data properly due to its enormous size [17]. Here are some of the privacy issues in Big Data as listed in Fig. 4.

1. Data Breach

A security incident in which the disclosure and access to sensitive and protected information by unauthorized users is considered a data breach. Data breaches can impact companies and their customers in a number of ways. A person can lose or steal

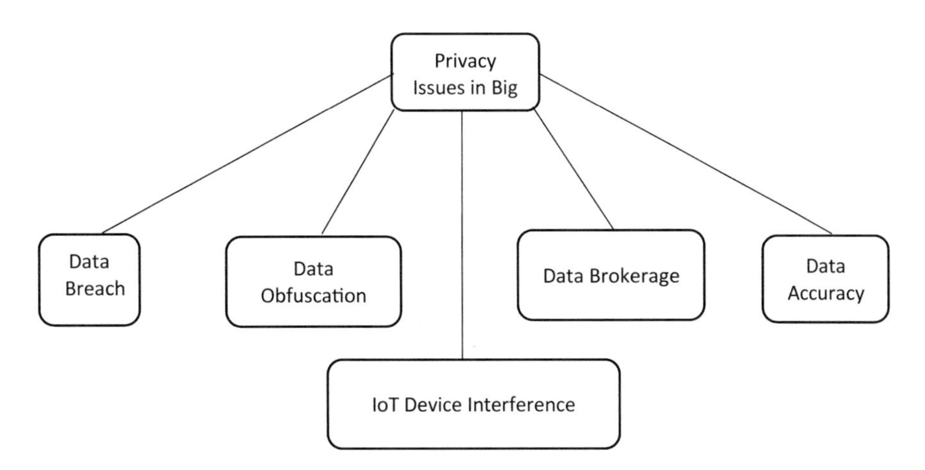

Fig. 4 Privacy issues in big data

the bank account, social security number, credit card numbers, email or passwords, etc., due to the data breach. Poor passwords, out-of-date applications, and targeted malware attacks are the result of infringements.

2. Data Obfuscation

In order to protect the data which is identified as personally identifiable information, the method of hiding the original data with changed content is known as data obfuscation [18]. Data obfuscation is often referred to as masking data. If not utilized properly, data obfuscation could result in complete failure, compromising the privacy of multiple individuals through big data analytics.

3. Data Brokerage

The process of collecting and selling the inconsistent and private data of the consumer to other businesses is called data brokerage. This is one of the serious privacy issues in big data as the data broker or information re-seller gather the private data of the customer from various sources like website cookies, reward cards, and social media and sold to other businesses companies in the form of profiles, who are looking to target people for different ad campaigns [19].

4. Data Accuracy

If the private data of an individual is incorrect and misplaced, then Big Data diagnosis for any organization does not work properly and can contribute to the lack of validation for data. As a result of which, it can harm the consumers since it results in the loss of jobs.

5. IoT Device Interference in Privacy

The use of IoT devices also causes big data privacy problems. IoT devices are the massive data collection engine of most different users' personal information. In addition, it should also be noted that by using information gathered about and from several devices related to a single person, an individual can become more readily recognizable and better identified.

2.3 *Privacy Strategies in Big Data*

Some of the strategies are available that the organizations can apply to maintain the data privacy. The strategies are as follows: -

1. Avoid Too Much Data Collection

For some time too much extra information causes a data protection failure. The social security numbers of their clients may not be needed by an agency or company; only customer IDs and passwords may be needed. Therefore, a company does not keep any extra consumer information that is completely unnecessary and leads to violating the customer's data privacy.

2. Implementation of Homomorphic Encryption

A form of encryption that allows users to compute data without decrypting it first is known as homomorphic encryption. The organizations should implement this form of encryption, to prevent the private information of customers from outside vendors.

3. Prevent from Internal Threats

Each and every organization today is facing certain threats related to privacy. Therefore, it is important to inform all staff about best practices to maintain data protection, such as logging off unused computers and other devices such as phones, laptops, etc., and regularly changing passwords.

4. Employ Real-Time Monitoring

Since a privacy problem can arise at any moment, companies should find a solution that can help to address privacy-related issues by tracking the data in real time. Organizations would thus be aware of the issue as soon as it arises and can take the necessary steps to fix it immediately.

2.4 Big Data Privacy Tool Features

Since we have discussed the strategies related to privacy in Big Data, let us discuss now some of the tool features and competencies an organization should look for.

1. User-Friendly Design

If business enterprises spend a lot of time to know and understand how to use the tools, then it is probably not a perfect match. In terms of design, the privacy tools must be user-friendly so that the workers can operate the tools quickly and can save time.

2. Automation

Many companies do not have adequate time to manually safeguard their consumer details from threats to privacy. Organizations should also choose a strategy that could help them to simplify the operation. Automation can free up businesses and help them achieve their goals and allow them to freely concentrate on running their business.

3. Cloud Compatible

The companies can easily accommodate customer requirements with the cloud service due to its scalability. That is why it is very essential for a big data privacy tool to be cloud compatible.

2.5 *Big Data Privacy Preservation Techniques*

To fight against the privacy threats, several privacy preservation techniques are available, which has given below: -

1. **K-anonymity**
2. **L-diversity**
3. **T-closeness**
4. **Randomization**
5. **Data distribution**

1. **K-anonymity**
   The method of extracting personally identifiable data from data sets so that the person's identity remains anonymous is called anonymization of data. Anonymization of data is one method of sanitization of information whose main purpose is to provide private data security. This is a model of privacy that can be used in a data-sharing situation to protect data from attacks. The final goal of this strategy is anonymity for the data subjects. If each published record on its QI attributes is indistinguishable from at least (K-1) others and at least 'K' is the cardinality of any query result on release data, a release of data is said to have the property of K-anonymity [19]. Via two generalization and suppression methods, the K-anonymity model decreases the granularity of data representation.

   Working of K-anonymity Method
   By taking a table as a database, consisting of **n** rows and **m** columns, the k-anonymization problem can be solved. Each row here represents a record that relates to a specific member of a population, and the various row entries in the table do not need to be identical. The values of the columns are the values of the characteristics associated with members of the population. Let us use a non-anonymized patient database to understand the K-anonymity process [9, 20].

   Example 1: Table 2 contains information about the patients like the name, gender, religion, state, and disease. The data of the record is not anonymized and to make anonymized, the techniques like suppression and generalization have been used applied.

   1. Suppression: The values of the attributes in this system are replaced by the asterisk '*' symbol. All or any of the values in a column can be replaced by '*'. The following table is anonymized by replacing all the values in the 'Name' attribute and all the values in the 'Religion' attribute with '*'.

   2. Generalization: In this method, each and every individual value of the attributes is replaced with a broader category. For example, the value '29' of the 'Age' attribute may be replaced by '≤ 30', the value '33' by '$30 < \text{Age} \leq 40$', etc.

Table 2 Non-anonymized patient database

Name	Gender	Age	Religion	State	Disease
A	Male	29	Hindu	Jharkhand	Corona
B	Male	27	Muslim	Kerala	Viral fever
C	male	39	Hindu	Kerala	Corona
D	male	29	Muslim	Odisha	Heart-related
E	male	33	Christian	Jharkhand	Corona
F	female	37	Christian	Tamil Nadu	Corona
G	female	38	Muslim	Tamil Nadu	Viral fever
H	male	34	Hindu	Odisha	Viral fever
I	female	27	Hindu	Kerala	Corona

Table 3 Anonymized patient database

Name	Gender	Age	Religion	State	Disease
*	Male	Age $\leq$ 30	*	Jharkhand	Corona
*	Male	Age $\leq$ 30	*	Kerala	Viral fever
*	Male	30 < Age $\leq$ 40	*	Kerala	Corona
*	Male	Age $\leq$30	*	Odisha	Heart-related
*	Male	30 < Age $\leq$ 40	*	Jharkhand	Corona
*	female	30 < Age $\leq$ 40	*	Tamil Nadu	Corona
*	female	30 < Age $\leq$ 40	*	Tamil Nadu	Viral fever
*	Male	30 < Age $\leq$ 40	*	Odisha	Viral fever
*	Female	Age $\leq$30	*	Kerala	Corona

In the database shown in Table 3, two anonymities are present with respect to the 'Age', 'State', and 'Gender' attributes. For any combination of these attributes found in any row of the table, there are always at least 2 rows with those exact attributes.

Example 2: Another example to understand the concept of K-anonymity. Table 4 is a database that consists of nine records and four attributes. In order to ensure three indistinguishable documents, the k-anonymity algorithm is used with a 'K' value around three. Anonymization on PIN and Age can now be implemented [15]. The following data will be included in the anonymized table. Here we have applied suppression and generalization in 'PIN' and 'Age' attributes and the corresponding result is shown in Table 5.

2. **L-diversity**

Table 4 Non-anonymized patient database

Serial no	PIN	Age	Disease
1	12,176	28	Corona
2	12,145	21	Corona
3	12,198	24	Corona
4	99,216	54	Viral fever
5	99,243	58	Corona
6	99,231	53	Viral fever
7	12,187	32	Corona
8	12,165	38	Viral fever
9	12,109	33	Viral fever

Table 5 Anonymized patient database

Serial no	PIN	Age	Disease
1	121**	2*	Corona
2	121**	2*	Corona
3	121**	2*	Corona
4	992**	>50	Viral fever
5	992**	>50	Corona
6	992**	>50	Viral fever
7	121**	3*	Corona
8	121**	3*	Viral fever
9	121**	3*	Viral fever

K-anonymity uses two generalization and suppression techniques to minimize the granularity of data representation. If the attacker presents context information and if the sensitive values of an equivalence class lack variety, K-anonymity fails to provide privacy, often called homogeneity attack. L-diversity is a strategy that decreases the granularity of data representation primarily in order to protect the data set's privacy. It is a type of anonymization which is based on a group. Let us take a look at Table 5.

If we know that Sohan is 24 years old and Sohan's pincode is 12176, then we can infer that even after anonymization, Sohan suffers from the corona virus, as shown in Table 5. This is called the attack of homogeneity. For instance, if Sohan is 38 years old and it is known that Sohan has no corona virus, then Sohan would certainly have a viral fever. This is called an assault of contextual information. Now, by using the following table, let us understand L-diversity.

It can easily be found from Table 6 that if we know that Sohan is 24 years old and Sohan's pincode is 12176, then Sohan is likely to be in the high-income

Table 6 Dataset with L-diversity

Serial No	PIN	Age	Disease	Salary (k)
1	121**	2*	Corona	70
2	121**	2*	Corona	80
3	121**	2*	Corona	81
4	992**	> 50	Viral fever	10
5	992**	> 50	Corona	7
6	992**	> 50	Viral fever	6

group because the incomes of all three people in the 121** pin are high relative to others in the table. This was seen as an assault on resemblance.

3. **T-closeness**
   It is a further enhancement of L-diversity, which is a group-based anonymization technique that decreases the granularity of data representation in order to protect data sets' privacy. The L-diversity model is generalized by treating the values of an attribute separately and taking into account the distribution of data values for that attribute. This can be done by the T-closeness privacy preservation model [21]. If the distribution of a sensitive attribute in the overall table is near to the distribution of a sensitive attribute in any similar class, then the class is said to have T-closeness property. In other words, an equivalence class is alleged to possess T-closeness if the gap between the distribution of a sensitive attribute during this class and also the distribution of the attribute in the whole table isn't any over a threshold 'T'. A table is assumed to have T-closeness if all equivalence classes have T-closeness [22].
   Let **R = *(r1, r2, ... rn)*** and **S = *(s1, s2, ... sn)*** are two the two distributions. Where.
   R = The distribution of a sensitive attribute in similar class.
   S = The distribution of a sensitive attribute in the overall table.
   So we can define the variation distance as:

$$\mathbf{VD}\,[\mathrm{R}, \mathrm{S}] = \frac{1}{2}\sum_{i}^{k}(r_i - S_i)$$

4. **Randomization**
   Randomization is the mechanism by which noise is usually applied to the data distribution of probabilities. In polls, sentiment analysis, etc., randomization is applied. Randomization does not necessarily require knowledge of other data records. During preprocessing and data collection time, it could be implemented. For randomization, there is no overhead anonymization. Nevertheless, it is not possible to apply randomization to large data sets due to time complexity and data utility.
5. **Data Distribution**

This technique helps to distribute the data in many sites. In two ways, data distribution can be performed.

1. Horizontal data distribution
2. Vertical data distribution

1. Horizontal Data Distribution

This kind of data distribution can only be implemented when such quantitative or aggregate functions or operations are implemented to the data without necessarily distributing the data. For example, if Big Bazaar needs to measure total revenue in different locations, they can use some tools that do statistical data computations. Nevertheless, as part of the data analysis, the data owner could need to share the data with the third-party vendor, which could lead to privacy violations.

2. Vertical Data Distribution

When the specific information of a person is spread through various websites under the jurisdiction of various organizations, then this kind of distribution is called vertical data distribution.

3 Security in Big Data

Big Data security is the shielding of data and analytics systems from any number of sources that might compromise their confidentiality both in the cloud and on the premises. The bulk of the tools focused on Big Data, and smart analytics are typically open-source. As a primary function, they are still not designed with defense in mind, leading to even more Big Data security issues.

3.1 Security Issues in Big Data

Some of the obvious Big Data protection problems that need to be tackled are shortlisted here.

- *Distributed framework*- Structures are distributed. Moreover, most Big Data applications spread enormous processing jobs across multiple networks for faster study. Hadoop is possibly the best-known instance of open-source software involved in this, which initially seems to have no safeguards of any kind. Distributed processing can mean that less data is managed by any single system, but it does mean many more systems where there may be security issues.
- *Endpoints*- Security solutions that retrieve logs from endpoints will have to check the authenticity of those endpoints; otherwise, the analysis will not generate any good result.

- *Entry checks*- As with IT enterprise as a whole, it is critically important to provide a system in which encrypted authentication/validation verifies that users are who they appear to be and determines who can see what.
- *Warehouse*- The data is usually stored at different levels in the big data architecture, based on the company's production vs. cost needs. For instance, high priority 'hot' data would usually be stored on flash media. A tier-conscious strategy would therefore mean locking down storage.
- *Data mining remedies*- These are the foundation of many big data environments; they detect patterns that signify company approaches. For that purpose, it is especially important to ensure that they are shielded from not only external threats but insiders who misuse network privileges to access confidential information, adding yet another layer of Big Data protection issues.

3.2 Security Challenges in Big Data

Besides the security issues, some of the challenges related to security are also present in Big Data and which are following: -

1. If the big data owner does not periodically update the system's security, they are at risk of losing and exposing data.
2. Without warning or permission, a Big Data supervisor may decide to mine data. If the intent is only for inquiry or illicit gain, the security systems need to monitor and report the suspicious access irrespective of where it originated.
3. The newest active-developing innovations are advanced analytical tools for unstructured big data. It can be hard for security tools and processes to secure such new toolsets.
4. The entry and storage of data are effectively protected by mature security equipment. However, from various analytics techniques to various sites, they may not have the same effect on data performance.
5. Troubles with cryptographic security. Although encryption is a very popular way of protecting sensitive data, it is also on our list of security challenges for big data. Given the possibility of encrypting big data and the essentiality of doing so, this security method is often ignored. Confidential information is usually stored in the cloud without cryptographic security.
6. Security assessments of big data help the organizations to become more cautious about their safety gaps. While it is advisable to carry them out on a daily basis, this recommendation is in fact rarely fulfilled.

3.3 Security Techniques for Big Data

Today, companies are increasingly collecting data according to the requirements of each department. However, the difficulty lies in the processing of this abundance of information, determining the key lessons, and then preserving the information. So, if the IT department and data scientists of an organization apply a variety of data analysis techniques to the gathered vast data, they should always guarantee that there is no risk of data leakage.

1. **Centralized Key Management**
   Virtual infrastructure control has been the standard way to maintain security for many years now. It applies equally strongly to regions with massive data, especially those with wide geographical distribution. Policy-driven automation, tracking, key distribution on demand, and the separation of key management from key usage provide best practices.

2. **Detection and Prevention of Intrusion**
   Intrusion detection and prevention systems are powerhouses for protection. That does not make them less useful for the Big Data platform. Big Data significance and distributed architecture lend themselves to attempted interference. Intrusion Prevention Systems (IPS) helps security administrators to safeguard against intrusion from the Big Data Network, and if an intrusion succeeds, Intrusion Detection Systems (IDS) will quarantine the intrusion until it causes serious damage [23].

3. **Physical Security**
   Don't neglect physical security. Construct it when you deploy your Big Data platform in your own data center or carefully apply the security of the data center of your cloud provider with due diligence. Physical protection systems may refuse outsiders or employees who do not have a company in sensitive areas entry into the data center. For video surveillance and security records, the same will happen.

4. **Encryption**
   In-transit and at-rest data needs to be secured by protection systems and vast volumes of data need to do so. User and machine-generated encryption must also work on many different data formats [24]. Encryption software would need to work with different analytics and output data software, as well as common large data storage formats such as relational database management systems (RDBMS), non-relational databases such as NoSQL, and advanced file systems such as Hadoop distributed file system (HDFS).

4 Conclusion

Big Data refers to complicated and broad collections of data that need to be collected and analyzed in order to discover useful data that can help corporations and companies. Privacy of Big Data requires the proper management of Big Data to reduce risk and protect confidential data. Different government and private sector organizations are constantly producing and gathering massive quantities of data, and it is very difficult to protect the data privacy of different users because of the immense scale of the data. Some of the threats to privacy in Big Data includes discrimination, disclosure, personal acceptance, and abuse and surveillance. The discrimination is a Inequality or bias which generally happens when some personal information of an individual is disclosed and whereas disclosure is an action of making new or secret information known. A leakage of personal information of a person can lead to personal embracement and abuse. Surveillance is the close observation of the private data to know the sentiment of o person. To handle these kinds of privacy threats, different privacy preservation techniques are available. The techniques are like T-closeness, randomization, and L-diversity. Big Data security is the processing of protecting and guarding data and analytic processes, both in cloud and on premise. Big Data heavily relies on the cloud but it is not the cloud alone that creates the security issues. The third-party applications can also create the security issues of the private data of a person. Some security issues are like distributed framework, endpoints, entry checks, data mining remedies. To solve these issues, certain technologies are available in the market like encryption, centralized key management, detection and prevention of intrusion, and physical security.

References

1. Sarkar D, Nath A (2014) Big data—a pilot study on scope and challenges. Int J Adv Res Comput Sci Manage Stud (IJARCSMS), 2(12):9–19(2014). ISSN: 2371-7782
2. Jena L, Kamila NK, Mishra S (2014) Privacy preserving distributed data mining with evolutionary computing. In: Proceedings of the international conference on frontiers of intelligent computing: theory and applications (FICTA) 2013. Springer, Cham, pp 259–267
3. Bhogal N, Jain S (2017) A review on big data security and handling. Int Res Based J 6(1). ISSN 2348-1943, March, 11, 2017
4. Gandomi A, Haider M (2015) Beyond the hype: big data concepts, methods, and analytics. Int J Inf Manage 35(2):137–144. https://doi.org/10.1016/j.ijinfomgt.2014.10.007
5. Wu X, Zhu X, Wu GQ, Ding W (2014) Data mining with big data. IEEE Transactions Knowledge Data Eng 26(1):97–107. https://doi.org/10.1109/TKDE.2013.109
6. Al-Kahtani MS (2017) Security and privacy in big data. Int J Comput Eng Inf Technol 9(2). E-ISSN 2412- 8856, February 2017
7. Arora M, Bahuguna H (2016) "Big data security—the big challenge. Int J Sci Eng Res 7(12). ISSN 2229-5518, December-2016
8. Pathrabe TV (2017) Survey on security issues of growing technology: big data. In: IJIRST, national conference on latest trends in networking and cyber security, March 2017
9. Mishra S, Tripathy N, Mishra BK, Mahanty C (2019) Analysis of security issues in cloud environment. Secur Des Cloud, Iot, and Soc Networking, pp 19–41
10. Mishra S, Mishra BK, Tripathy HK, Dutta A (2020) Analysis of the role and scope of big data analytics with IoT in health care domain. In: Handbook of data science approaches for biomedical engineering. Academic Press, pp 1–23
11. Sagiroglu S, Sinanc D (2013) Big data: a review. In: Proceedings of the 2013 international conference on collaboration technologies and systems (CTS), San Diego, CA, USA, 20–24 May 2013; pp 42–47

12. Hashem IAT, Yaqoob I, Anuar NB, Mokhtar S, Gani A, Ullah Khan S (2015) The rise of "big data" on cloud computing: review and open research issues. Inf Syst 47:98–115
13. Mishra S, Tripathy HK, Mishra BK, Sahoo S (2018) Usage and analysis of big data in e-health domain. In: Big data management and the internet of things for improved health systems, IGI Global, pp 230–242
14. Kalaivani R (2017) Security perspectives on deployment of big data using cloud: a survey. Int J Adv Network Appl (IJANA), 08(05):5–9
15. Rao PR, Krishna SM, Kumar AS (2018) Privacy preservation techniques in big data analytics: a survey. J Big Data 5(33):1–12. https://doi.org/10.1186/s40537-018-0141-8
16. Kumaresan A (2015) Framework for building a big data platform for publishing industry. In: Knowledge management in organizations. Springer International Publishing, Cham, Switzerland, pp 377–388
17. Wei G, Shao J, Xiang Y, Zhu P, Lu R (2015) Obtain confidentiality or/and authenticity in big data by ID-based generalized signcryption. Inf Sci 318:111–122
18. Kuzu M, Islam MS, Kantarcioglu M (2015) Distributed search over encrypted big data. In: Proceedings of the 5th ACM conference on data and application security and privacy, San Antonio, TX, USA, 2–4 March 2015; pp 271–278
19. Tan Z, Nagar UT, He X, Nanda P, Liu RP, Wang S, Hu J (2014) Enhancing big data security with collaborative intrusion detection. IEEE Cloud Comput 1:27–33
20. Karle T, Vora D (2017) PRIVACY preservation in big data using anonymization techniques. In: 2017 International conference on data management, analytics and innovation (ICDMAI) Feb 24, IEEE, pp 340–343. https://doi.org/10.1109/ICDMAI.2017.8073538
21. Jain P, Gyanchandani M, Khare N (2016) Big data privacy: a technological perspective and review. J Big Data 3:25. https://doi.org/10.1186/s40537-016-0059-y
22. Li N, Li T, Venkatasubramanian S (2007) t-closeness: privacy beyond k-anonymity and l-diversity. In: 2007 IEEE 23rd international conference on data engineering, Apr 15, IEEE, pp 106–115. https://doi.org/10.1109/ICDE.2007.367856
23. Cardenas AA, Manadhata PK, Rajan SP (2013) Big data analytics for security. IEEE Secur Priv 11(6):74–76. https://doi.org/10.1109/MSP.2013.138
24. Gai K, Qiu M, Zhao H (2017) Privacy-preserving data encryption strategy for big data in mobile cloud computing. IEEE Trans Big Data. https://doi.org/10.1109/TBDATA.2017.2705807

Chapter 5
Comparative Analysis of Anonymization Techniques

Arijit Dutta, Akash Bhattacharyya, and Arghyadeep Sen

1 Introduction

Companies such as healthcare, retail, digital media, and financial industries await massively on data with information that are personally identifiable (PII-Personally Identifiable Information). This enables advances in technology and research. If your personal data works well, it can fall into the amiss grips threatening the safety of thousands and thousands of people around the world. The vast amount of information available today and the increasing computation capacity accessible to attackers cause these ambushes a deliberate issue [1]. Cybersecurity is on the rise. Cybercriminals are expected to steal 144 billion records by 2023. State-specific regulations such as GDPR, HPAA, PCI DSS, and California-specific consumer privacy laws have appeared and have been developed to orate these flourishing security challenges. This means that the methods and protocols for data collection must comply with these rules, and operational service providers (MSPs) must assist customers to safeguard proper and secure data collection is implemented.

As information technology advances rapidly, businesses are now collecting, and large extent of information is stored in databases. Mostly, this information is hoarded as a table, and each record is personalized. Each record has several properties that can be divided into three properties.

1. A clear identifier that can clearly identify an individual.
2. Identify traits that can be easily identified by an individual with an identity.
3. Sensitive features that are considered sensitive and should not be exposed [2].

Several various anonymization methods have been studied to insulate the accused identity. Other data holders often delete or encrypt explicit identifiers. In order to anonymize information that does not administer anonymity, the information disclosed also includes auxiliary data called QCI identifiers that can be employed to

A. Dutta (✉) · A. Bhattacharyya · A. Sen
School of Computer Engineering, KIIT Deemed to be University, Bhubaneswar, India

P. K. Das et al. (eds.), *Privacy and Security Issues in Big Data*, Services and Business Process Reengineering, https://doi.org/10.1007/978-981-16-1007-3_5

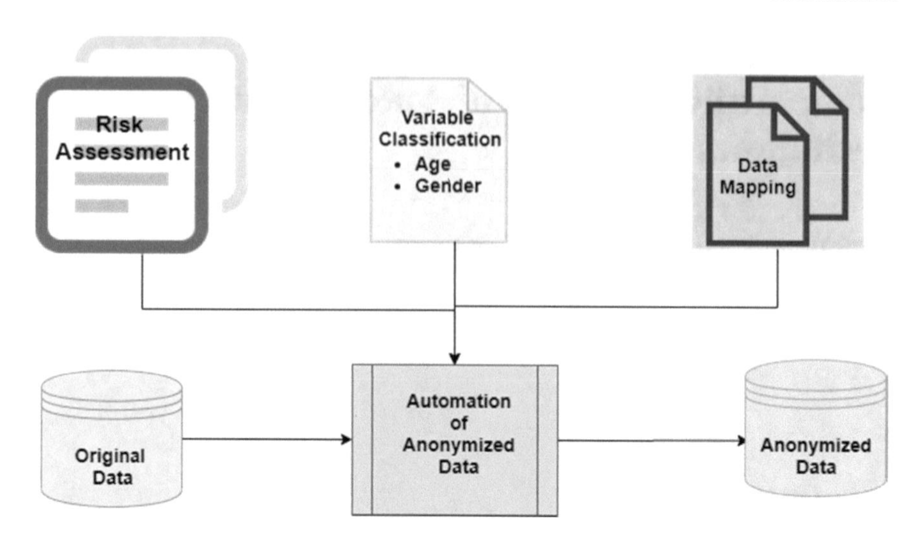

Fig. 1 Data anonymization process

re-identify data respondents so that information that does not need to be disclosed can be disclosed. When disclosing information, it is essential not to disclose personal sensitive information. Published tables provide resourceful information to the researchers, while likewise providing exposure defy to individuals with data in the tables. Therefore, the main goal is to curb the hazard of exposure to an appropriate level. This can be accomplished by anonymizing the data prior to it becomes public. To effectively limit exposure, you need to assess the risk of exposure of anonymous tables (Fig. 1).

2 Anonymization of Data

Data anonymity is the mechanism of assuring personal or delicate material by encrypting or erasing identifiers that link individuals to data that is stored [3]. Data anonymity means that identifiable information such as age, gender, name has been replaced or evacuated from the data set, making it preposterous to state whether an entity has been exposed to the data. You have to make a decision [4]. This is generally cited as “data hygiene” or “data masking”.

For example, PII like name, social security number, and address may be enforced through an anonymous process that manages data but keeps the source anonymous. However, erasing the data on the identifier could allow an attacker to restore the data anonymization process using de-anonymization techniques. Since data typically goes through multiple sources, you can use some public anonymization methods that allow you to bypass the sources and disclose personal information [5].

Industries that depend on the accumulation of sensitive personal data should adopt a few forms of data anonymity. The degree of anonymity depends on the variety of the business, the kind of data collected, and even if a group of recipients, called a controlled release, is shared publicly or privately. Some elements of the data have not been compromised to provide value, but the data must be anonymous enough so that the information is not available to hackers in the event of a breach [6]. There is no direct personal identifier for data that is reasonably unknown, such as name, address, social security number, or phone number. This may include indirect identifiers that can be combined to identify an individual, including work, salary, or diagnosis. Nicknames are used to protect sensitive personal information similar to anonymity. However, nicknames do not remove all identifying information, and it is difficult if not impossible to identify a person on the left.

3 Prevailing Illustrations of Data Anonymization

Medical Research: Medical professionals and researchers would like to examine the data on the prevalence of certain diseases in some demographics. It protects patient privacy and meets HIPPA standards.

Promoting Marketing Reforms: Many online retailers want to discuss with their customers when and how to do this via email, social media, digital advertising and Web sites. In order to improve their services and meet the growing demand for positive or unique customer experiences, digital agencies depend on insights derived from user data. In order to search for relevant information while halting, these marketers and analysts should take advantage of data anonymity [7].

Software and development of the product: Developers often depend deeply on virtual data to improve efficiency, iron out new challenges, and develop new tools that can improve service delivery. This kind of data should be anonymous so your personal information is not at high risk in case of the data breach [8].

Performance Business Performance: Multiple large companies collect employee-related information to amend performance, increase productivity, and improve the safety of the employees. Data anonymity and aggregation allows these organizations to obtain important information without having to diagnose, monitor, or misuse their employees.

4 Data Anonymization Techniques

The process of Data Masking is to hide data with modified values. You can create a displayed version of the database and use conversions such as pushing, encryption, entering a name, or retrieving a letter. For example, you can change the value character to a symbol such as “*” or “x”. Encrypting data makes back-end engineering or detection impossible (Fig. 2).

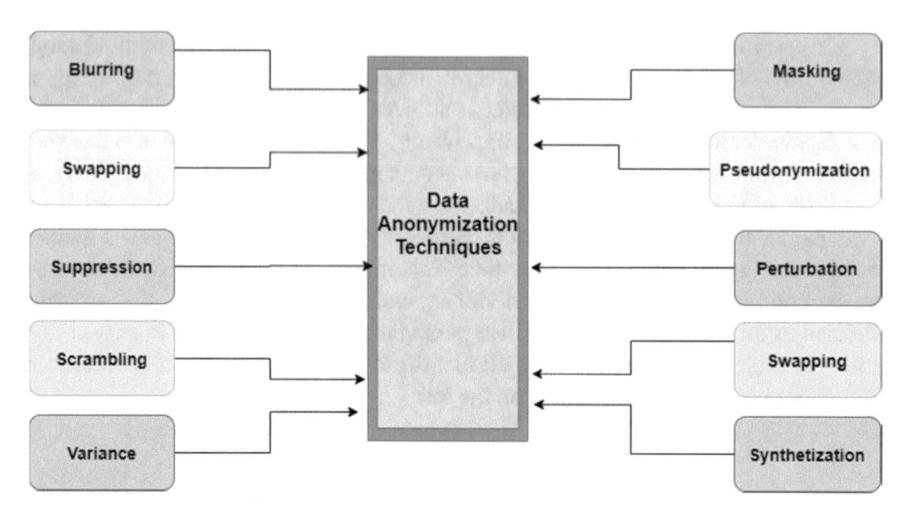

Fig. 2 Common anonymization techniques

Impersonation: Data mimicry management and identification method instead of identifying your identities with duplicate identifiers or pseudonyms (e.g., "John Smith" identifiers are replaced by "Mark Spencer"). Aliases allow the use of modified data for statistical accuracy and data integrity. Keep it. Training, development, testing and analysis while protecting data privacy.

Normalization: Some data is intentionally deleted and difficult to detect. Data can be organized in a set of parameters or a large area with appropriate parameters. You can remove the house number from the address, but do not delete the street name. Its desire is to remove certain identifiers while maintaining a certain level of data accuracy.

Data Transformation: It is also known as push-through and retrieval, is a method used to sequence data values so that they do not follow the original record. For example, converting an asset that contains an identifier as date of birth (column) could have a greater impact on anonymity than membership type values.

Stop Data and Set up Data: Modify original data set by adding round numbers and random audio to the original data set. The width of the values should be proportional to the damage. The smaller the base, the weaker the anonymity, and the larger the base, the less data usage. For example, rounded values such as age or house number are equal to the actual value, so you can use 5 as a base. If you multiply the house number by 15, the value is reliable. However, as you use a higher condition, such as 15, you may have duplicate age values.

Nt Synthetic Data: Data transaction data for Nt is generated by an algorithm that is not related to the actual event. User data sets are used to create artificial data sets instead of altering actual data or threatening privacy and security. This process involves creating a statistical model based on the model found in the original data set. You can use standard deviation, intermediate, precise order, or other mathematical methods to generate synthetic data.

And Price and Date Deviations: Algorithms can be handled to convert numerical data values to random numbers. These small steps can form a huge difference if they are used properly.

Mixed Scratching: In proper scrubbing, the characters are combined and rearranged, making it impossible to see the actual data. A simple example of this is the change of the "Daniel" name to "Leniod" and the "Jacqueline" to "Kalinizu".

Numerical Opacity: Instead of completely masking the ambiguous data values, they are altered to move away from the actual values, protecting the identity of the individual. Numerical ambiguity can be done in a variety of ways, including reporting rounded values or group averages.

Ress Suppression: In some instances, data columns and/or records might remain in the data set, which may not be helpful in the evaluation of the data, but the information can be identified. In such instances, it is best to suppress or delete columns and or records. It is very much important to whole delete the data from the spreadsheet, not hide it (Tables 1 and 2).

In this example, values for all attributes have been swapped:

1. Risks associated with Data Anonymization

It leads to tremendous progress for companies in the field of data computing, but it is not without limits and risks. Improperly implemented or weak algorithms can lead to false anonymization.

Table 1 Prior to anonymization

ID	Name	Acc type	Subscribed date	Ticket submitted
1	Luke	Pro	13 April	3
2	John	Org	25 Feb	3
3	Nathan	Free	17 Sep	4
4	Aaron	Free	2 May	1
5	Daniel	Pro	14 Aug	2
6	Michael	Pro	18 Dec	0

Table 2 Post anonymization

ID	Name	Acc type	Subscribed date	Ticket submitted
1	Daniel	Pro	14 Aug	2
2	Nathan	Free	17 Sep	4
3	Michael	Pro	18 Dec	0
4	Luke	Pro	13 April	3
5	Aaron	Free	2 May	1
6	John	Org	25 Feb	3

Identity Disclosure: The term disclosure of identity, also known as single out, is used to describe a situation in which everyone or some people can be identified in a data set.

Attribute Disclosure: Attribute disclosure is the ability to determine whether a specific individual has placed an attribute in a data set. For example, an unknown data set could show that all employees in the sales department of a particular office arrived after 10 a.m. If you know that certain employees are in the sales department of this office, they will arrive after 10 a.m., although the specific ID is in the data set.

Possible Connectivity: Connectivity refers to the ability to add two or more data points to a single data set or another to create a more harmonious image of a particular person.

When the injection is exposed: The injection is exposed when you are confident about the value of your symptoms based on other symptoms.

Proper data anonymity can be time-consuming and difficult, but experienced professionals must keep these practices strictly in order for businesses to stay compliant and avoid attackers. There are many data anonymity tools and software out there that can help businesses overcome data anonymity barriers and gain the many benefits of data collection.

2. Classification of Anonymization Techniques

Data anonymity is the method of applying data to ensure personal identification, allowing data to be shared and analyzed securely. Figure 3 depicts the analysis of various anonymization methods and the used algorithms.

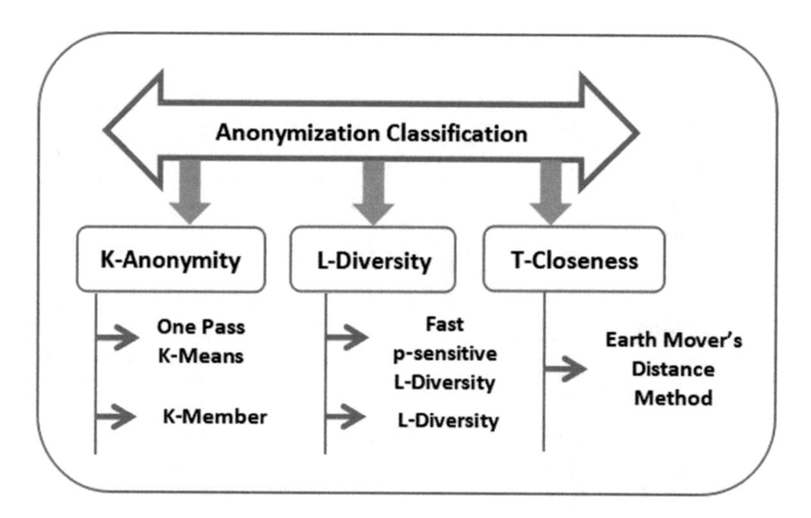

Fig. 3 Classification of anonymization

K-Anonymity

Sweeney [9] introduced the K-Declaration as the non-transferable property of at least one K-1 record for the i-Identifier. A simple anonymous approach is the introduction of normalization at the cell level. June-Lin Lin et al. [10]. An effective clustering method is used for anonymization. The algorithm exercised is a pass. The algorithm is partitioned into two stages: a clustering stage and an integration stage. In [11], authors proposed a technique that uses clustering thinking to minimize loss of data and ensure better data quality. In this presented paper, we use the greedy K-member clustering algorithm to verify effectiveness based on data quality, performance, and scaling. Secret: Useful full domain K-anonymity proposed by Raghu Ramakrishnan et al. [12] provides a practical framework for implementing the entire DOM.

This is a simple modification using the kernel-based K-object clustering approximation proposed by Tripathi et al. [13]. The various clustering methods in this paper are considered to be one of the first methods of dealing with uncertainty in blurred mines. Krishnapuram and Keller [14] proposed a possible hybrid approach to clustering for privacy in application domain. According to Ciriani et al. [15], K-anonymity is divided into two modes: generalization and prevention. Xiao et al. [16] proposed the concept of personal anonymity, which created minimal generalization and contained large amounts of data.

L-Diversity

If the sensitive trait has at least an L "well-advanced" value, the equivalent class is called L-variant. If there is an L-variable for each class in the table, it is said that the table has an L-variable [17]. Ashwin Makhanwazla et al. [18] identified two attacks occurring in K-anonymity. The authors say that for Tripathy et al. sensitive L-variation anonymity algorithm, the definition of the proposed novel is L-variation, and the q-block is L-variation when the sensitive property has an L "well expressed" value. The third step, called L-variable algorithm to achieve L-variable, is combined into two steps of OKA.

T-Closeness

The equivalent class is called T-Clogeness if the distance amongst the sensitive attribute distribution of this class and the attribute distribution throughout the table does not exceed the limit. The necessity of T-intimacy is that the distribution of sensitive properties in Eq. Classes is near to the distribution of sensitive features throughout the table. T-Proximity: The privacy coupling proposed in [17] accounts for K-anonymous and L-variation using the earth motion distance calculated using the distance amongst the prior necessities. The EMD is calculated for two attributes: a numerical attribute and a hierarchical attribute. Luo Yongcheng et al. [18] conferred that guarding data privacy is a crucial issue when accumulating micro-data. K-anonymity could prevent link attacks, diversity can prevent homogeneous attacks, and T-proximity can reduce information loss. Similarly, there are several other research works undertaken in the field of security and privacy issues in various application domains.

5 Comparative Analysis

In this section, we used these methods to conduct a comparative analysis based on privacy criteria, i.e., data loss. Many real data sets are used to compare personal information. The data set includes a transportation system data set, a census marriage data set, and criminal status.

1. Transmit data set (size: 465): IMEI number, latitude, longitude, X, Y, Z values contain accelerometer and gyroscope, IMEI number is a sensitive function, the rest is partially detected. Figure shows the information loss analysis using transportation data samples.
2. Wedding gift (size: 2348): There are five characteristics: Year, Age, Marital Status, Gender and people out of which age is a sensitive attribute and rest of them are quasi-identifying. (Fig. 4).
3. Crime data set by country (size: 16,422): There are five attributes: country, crime type, crime, year and number, of which the country acts as a sensitive function and the rest are partially detected. Figure 5 depicts the crime data set and data loss associated with it.

The figures above reflect the analogy of different Benami methods, i.e., K-anonymity, L-mutation, and T-clone, for different data sets based on privacy metric data loss where K, L, and T represent values. Anonymous feature. no. As symptoms increase, data loss increases, indicating that data loss is directly proportional to the number. Anonymous feature. So, from the above comparison, we conclude that T-proximity has comparatively low data loss than the L-variation and K-anonymity. A comparative analysis of various anonymity techniques is highlighted in Table 3.

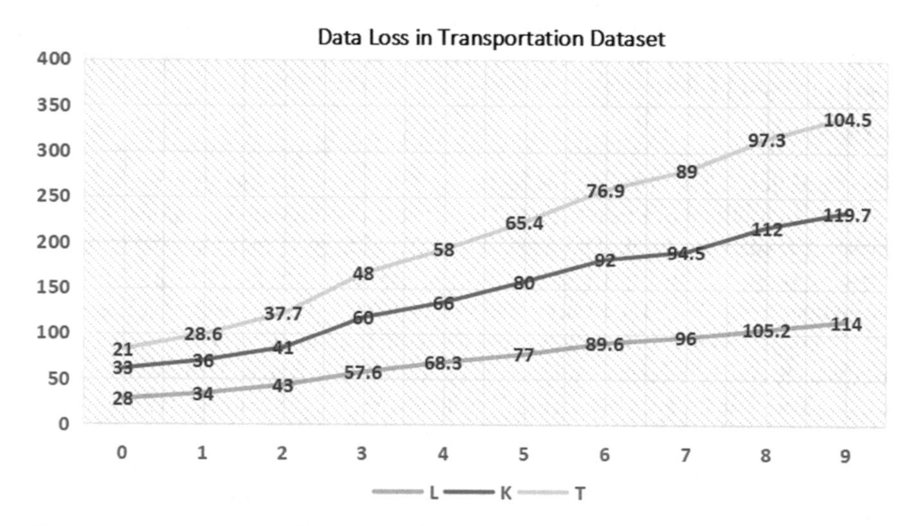

Fig. 4 Data loss in transportation data samples

Fig. 5 Data loss in crime data samples

Table 3 Comparative analysis of anonymity techniques

Criteria	k-Anonymity	X-Y-Anonymity	L-Diversity	T-Closeness
Model	Record linkage	Record linkage	Attribute linkage	Attribute linkage
Applied with numerical	Yes	Yes	Yes	No
Data utilization	Good	Average	Average	Poor
Sensitive feature support	Yes	Yes	Yes	No
Privacy degree	1	2	3	4
Skewness	Yes	Yes	Yes	Yes
Threshold requirement	No	Yes	Yes/No	Yes
Similarity attack	Yes	Yes	Yes	No

6 Conclusion and Future Work

It is clear in distinction to the literature today that consumer privacy is a major concern. Various models stated for micro-data have been embraced to protect the confidentiality of various types of data such as transportation system data, medical data, and marriage census data. This paper focuses to review various anonymization methods such as K-anonymous, L-variant, and T-almost. The use of these methods can be evaluated for a wide variety of data sets, and T-clauses have less data loss than L-variation and K-anonymity, although these methods still lead to extensive data loss. Therefore, advanced technologies have the potential to reduce data loss and provide privacy when making better use of publicly available data. In future work, we will correlate these methods with other matrices.

References

1. Chen B-C, Kifer D, LeFevre K, Machanavajjhala A (2009) Privacy-preserving data publishing. Found Trends Databases 2(1–2):1–167
2. Li N, Li T, Venkatasubramanian S (2007) T-Closeness: privacy beyond kAnonymity and L-Diversity. In 23rd IEEE International Conference on Data Engineering. IEEE Press, Istanbul, pp 106–115
3. Shanthi AS, Karthikeyan M (2012) A review on privacy preserving data mining. In: IEEE International Conference on Computational Intelligence and Computing
4. Wang P (2010) Survey on privacy preserving data mining. Int J Digital Content Technol its Appl 4
5. Kedar S, Dhawale S, Vaibhav W, Kadam P, Wani S, Ingale P (2013) Privacy preserving data mining. Int J Adv Res Comput Commun Eng 2(4)
6. Sweeney L (2002) Achieving k-anonymity privacy protection using generalization and suppression. Int J Uncertainty, Fuzziness Knowl-Based Syst 10(5):571–588
7. Lin J, Wei M (2008) An efficient clustering method for k-anonymization. In: Proceedings of the 2008 international workshop on privacy and anonymity in information society .ACM, New York, pp 26–35
8. Byun J, Kamra A, Bertino E, Li N (2007) Efficient k-anonymization using clustering techniques. In: Kotagiri R, Krishna P, Mohania M, Nantajeewarawat E (eds) Advances in databases: concepts, systems and applications, vol 4443. LNCS. Springer, Heidelberg, pp 188–200
9. LeFevre K, DeWitt D, Ramakrishnan R (2005) Incognito: efficient full-domain K-anonymity. In: International conference on management of data. ACM, New York, pp 49–60
10. Tripathy B, Ghosh A, Panda G (2012) Kernel based K-means clustering using rough set. In: International conference on computer communication and informatics. IEEE Press, Coimbatore, pp 1–5
11. Data Anonymization. http://www.slideshare.net/KaiX/lions-zebras-and-bigdata-anonymization
12. l-diversity and t-closeness. http://www.utdallas.edu/~muratk/courses/dbsec12f_files/DBSec_priv3.pdf
13. Ciriani V, Vimercati V, Foresti S, Samarati P (2007) k-Anonymity. In: Kikuchi H, Rannenberg K (eds) Advances in information and computer security. LNCS, vol 4752. Springer, Heidelberg
14. Xiao X, Tao Y (2006) Personalized privacy preservation. In: International conference on management of data. ACM, New York, pp 229–240
15. Li N, Li T, Venkatasubramanian S (2007) T-Closeness: privacy beyond kAnonymity and L-Diversity. In: 23rd IEEE international conference on data engineering. IEEE Press, Istanbul, pp 106–115
16. Machanavajjhala A, Gehrke J, Kifer D (2006) L-diversity: privacy beyond KAnonymity. In: 22nd international conference on data engineering. IEEE Press, Atlanta, GA, USA, pp 24
17. Tripathy B, Maity A, Ranajit B, Chowdhuri D (2011) A fast p-sensitive l-diversity Anonymisation algorithm. In: Recent advances in intelligent computational systems (RAICS). IEEE Press, Trivandrum, pp 741–744
18. Yongcheng L, Jiajin L, Jian W (2009) Survey of anonymity techniques for privacy preserving. In: International symposium on computing, communication, and control. IACSIT Press, Singapore, pp 248–252

Chapter 6
Standardization of Big Data and Its Policies

Sankalp Nayak, Anuttam Dash, and Subhashree Swain

1 Introduction

Complex and large data that is not possible to be processed by a traditional data processing system is otherwise known as big data [1, 2]. It consists of structured, unstructured, and semi-structured data that are used in a particular industry of domain frame. The amount of data generated by healthcare applications, IT firms, sensor networks, and the Internet are rising day after day due to the rapid development of technology. All these heavy amounts of data produced from disparate data sources at quite a high velocity [3] are collectively hereinafter as big data. It [4, 5] can be identified as "a new generation of architectures as well as technologies, built in order to be isolated economically benefit from a very big quantity of different forms of data, by ensuring high-speed data fascinate, analysis as well as discovery". There are primarily three characteristics of big data which are otherwise called 3Vs, i.e., velocity, volume, and variety. Later, through studies, it was found that these 3Vs are not sufficient to explain the true sense of big data that we face. Thus, some other Vs have been added and they are [6] value, veracity, validity, venue, vocabulary, vagueness, and variability. Big data environment involves diverse forms of data which might be audio, image video, or just XML, JSON files, or spreadsheets. This phenomenon is justified by the variety of components of big data. To ensure privacy in the big data environment, diverse methods have been adopted in the current years. These processes can be clustered into different stages of the data life cycle [7]. Figure 1, for example, processing, generation of data, storage. In order to ensure privacy protection in the data generation phase, falsifying data techniques as well as access restriction techniques are used. Different encryption techniques ensure privacy safeguards during the data management process. Encryption methods could be additionally grouped into storage paths, attribute-based, and identity-based encryption.

S. Nayak (✉) · A. Dash · S. Swain
School of Computer Engineering, Kalinga Institute of Industrial Technology (Deemed to be) University, Bhubaneswar, India

P. K. Das et al. (eds.), *Privacy and Security Issues in Big Data*, Services and Business Process Reengineering, https://doi.org/10.1007/978-981-16-1007-3_6

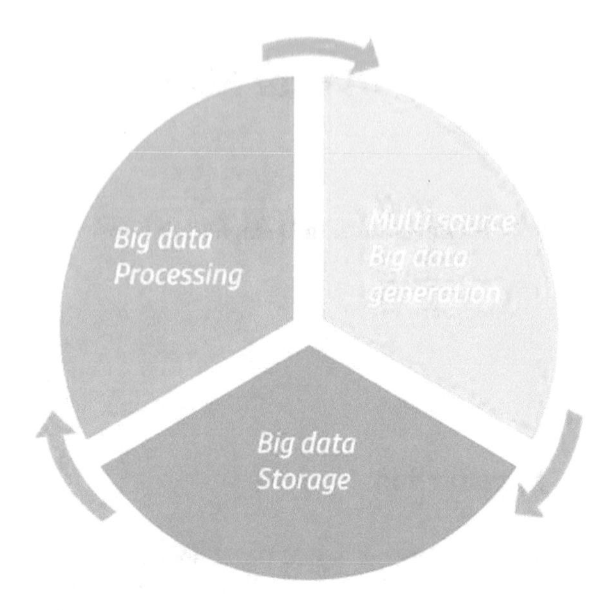

Fig. 1 Big data life cycle phases

Hybrid clouds which mainly consist of private clouds are used in order to store very sensitive information. Knowledge extraction from data and privacy-preserving data publishing (PPDP) are incorporated in the data processing phase. In PPDP, several methods like suppression as well as generalization are used to protect the integrity of data. Such processes can be furthermore grouped within association rule, clustering, a classification which are forms of data-mining techniques. While data is being divided into distinct groups in classification and clustering, association rule mining-based techniques are used to detect the relationships, structures, and patterns in the data [8]. To handle the high-speed data which is expensive in nature, an efficient and effective framework of big data is required. Big data should go through several stages in its entire life.

According to the study done in 2012, where it was found that on average, every day around 2.5 quintillion data bytes is produced. The rate at which data is generated is very fast and the volume is vast [9]. Algorithms that are robust, accurate, require minimum pre-processing requirements, and are lightweight in nature should be considered. Like, in data mining, lightweight swarm search and expedite PSO can be used for feature selection [10]. Further, the Internet of things has enabled the interconnection of any kind of system that humans care about that in turn produces a very big amount of data at every instance in a day [11]. In fact, the IoT environment is one of the key triggers of big data analysis.

Over the past few days, the heavy storage of vast amounts of data in big data environments has led to big data analytics which can be used to extract very useful information about health care and other problems of our society. Smart energy analytics are extremely challenging and complex in nature which also part in many numerous problems with the general form of big data analytics. In order to ensure safe operation in real-time systems, data intelligence is used in big data smart energy environments [12]. This in turn can be used to create better business opportunities for the companies and also marketing purposes. As the databases have very sensitive information like the personal attributes of an individual, they cannot be allowed direct access for analysts and researchers. In this case, the privacy of an individual is disclosed and therefore is considered illegal. Big data privacy is essential in the data collection, the storing of data, and the production of data in the lifecycle of data. Privacy is the opulence of jurisdiction for the storage and use of personal knowledge. Awareness of privacy is the right of a person or group to forbid the dissemination of knowledge about themselves to persons other than those to whom information is given. One big concern for individual privacy is the identification of individual privacy [13].

2 Conventional Data Privacy Standards Versus Big Data Privacy

We already know that data privacy is not a very new topic to talk about. It was a recognized issue for financial and healthcare information by the late 1970s. In those days, the data privacy principles adopted for data privacy protection were referred to as "Fair Information Practices" (FIP).

There were mainly five tenants under FIP:

- Openness: There should be no secret collection of personal data by any form of a system.

Disclosure

- Organizations should clearly mention before the users what information they collect and how they use that information.
- Secondary usage: Without authorization from an individual, the data gleaned for a particular reason cannot be used in some other context.
- Correction: Individuals should be empowered by the right to correct or modify any form of their personal data.
- Security: Any organization which primarily deals with the using, disseminating, maintaining, and creating personal data, should assure data privacy and should ensure proper precautions taken to check data misuse.

With the evolution in the data domain, there is a new focus in the data source which is otherwise referred to as lineage. It has become necessary to comprehend correctness and the overall standard of the data and the use of the data in correlation with industry compliance criteria and personal privacy. It leads to more complexity with the data becoming an advantage for both the users and the company.

3 The Drive for Self-Service and the Necessity for the Protection of Big Data

3.1 Privacy and Security Issues

This is a key topic in big data. Big data's protection paradigm is inappropriate in relation to complicated systems and because of this, it is diminished by nature. That being said, with its deficiency, data can still be easily compromised. Big data privacy issues must be balanced with the important and primary goals of business just like any other security or privacy issue. That data is more easily accessed by the customer in online customer experiences in e-commerce platforms. Therefore, the data have a much more direct influence on the information at the bottom. Moreover, because of the lack of a user-friendly and effective customer experience, the customer base can shift to other sites. Similarly, a poor Web-based support system can lead to increased swirl. There are some risk factors involved in this kind of "transparency". More interactive operations like self-service operations with customers mean you are gathering more and extra data about them which might be sensitive such as their preferences, purchases, and their accounts. More data can improve the experience but it might lead you to a potential risk situation. There is a risk of confidential and extremely personal information to be exposed. As a result with the organizations collecting data for public consumption, there has been a driving force for improvement in the IT Governance and data privacy efforts. And hence, now both the IT and business are collectively realizing that data protection by ensuring data privacy is everyone's responsibility.

Privacy information is a right in order to get an authority on how information is obtained and then utilized. Data privacy is the right of a person or a community to prevent information about themselves from being made available to individuals other than those to which they offer access. A big concern with respect to the privacy of users is the collection of sensitive information through data transmission [13]. Data protection focuses on the use and regulation of user data things such as the establishment of policies in order to ensure that customer-sensitive information is stored, exchanged, and processed in an acceptable manner.

Security is the task of protecting information and information resources by using technologies, procedures as well as preparation based on—Unlawful entry, dissemination, interruption, alteration, review, monitoring, and destruction. Security focuses mainly on the protection of data from phishing software and exploitation of stolen data to make a profit [14].

3.2 Privacy Standards in a Big Data World

Big data analytics are utilized by several organizations; a substantial majority among themselves prefer not to use these technologies because of a significant lack of basic privacy and security software. These articles discuss potential solutions to improve big data systems with the aid of privacy security aspects. Concepts and development approaches of the framework that support:

- concretization of the privacy policy for governing access to data gathered on appropriate big data platforms
- Generating effective compliance controls for such guidelines
- That incorporation of the produced monitors into the targeted analytical platforms. The compliance strategies suggested for conventional DBMS priority is not appropriate for large data set-up due to the rigid deployment requirements needed for huge data sizes, data volume, and data processing speed.
  When designing big data privacy efforts, the first point which can help you is the data source and how the data is going to be used. Businesses and government departments produce and continually accumulate vast volumes of data. The present intensified emphasis on large volumes of data would inevitably generate challenges and possibilities for understanding the analysis of those data across a wide variety of realms. However, the power of big data comes at a cost; the safety of customers is always at stake. Making sure compliance with privacy policies and regulation is restricted in emerging big data analytics and mining activities. Therefore, the conversation can be driven in the way of how we are meant to or we are not compelled to cause the exploitation of data. Nevertheless, there is an attitude to evade the extremely important topic of data protection and data privacy support in this digital world. The factor which needs to be addressed is how to ensure a proper balance between:
- The value for an individual or end-user.
- The privacy and protection levels necessary for your customer and you.

These practices have to be followed if you want to see the business growing rightly and credible for your customers.

To overcome these problems, recognize the desire toward new approaches to informal methodology, and procedures in testing. Latest privacy conformance evaluation paradigms for quadruple areas in Extract, Transform, and Load (ETL) method as seen in Fig. 2 [15, 16] as well.

1. *Validation of the pre-Hadoop process*: The following stage represents a phase of data loading. At this point, privacy standards describe sensitive data objects which are specifically identified by the individual or institution. Privacy terms may also specify which data parts must be retained and for how long. Schema restrictions may also occur at this point.
2. *Map-reduce method validation*: In this procedure, it alters the big data resources to respond efficiently to an inquiry. Privacy parameters stipulate the least amount of retrieved documents which demand to preserve data of individuals.
3. *Validation of ETL process*: Related to phase (2), the warehousing justification for complying with the privacy requirements should be verified at this point. Any values of data can be summarized secretly or omitted in the warehouse as this implies a high risk of recognizing persons.

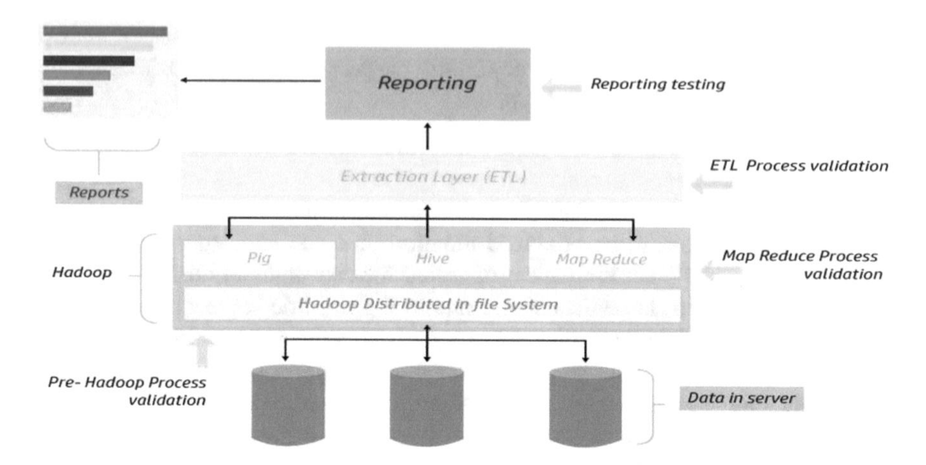

Fig. 2 Privacy compliance monitoring for the quadruple areas of Extract, Transform, and Load (ETL)

4. *Testing the results*: Reporting is an alternative kind of inquiry, possibly with greater visibility and a broader audience. Privacy words that describe 'intent' are important for ensuring that confidential data shall not be disclosed except for defined purposes.

3.3 Privacy of Big Data in the Data Production Process

The two categories in the generation of data are passive and active data generation. The generation of active data means the data owner can send data to a third-party [17], whereas the generation of passive data applies to the conditions in which the data is generated by the data owner's Internet practices and the owner of the data does not know that the data may be accessed by an external source. Minimizing the risk of privacy violation in the processing of data, possibly by restricting access.

1. Accessibility limitation: Indeed the owner of data feels that the data will be exposed, is classified information that is not to be exchanged, he or she declines to give that data. Unless the data owner offers the data voluntarily, a few steps may be provided in order to ensure that privacy, like anti-tracking plugins, ads or script blocks, and encryption methods.
2. Falsification data: Under certain situations, it is impossible to deny access to confidential data. In any scenario, data could be distorted by the use of the methods subsequent to the data obtained from a third-party. In case the data is corrupted, the real information cannot be readily exposed. The data owner uses the following techniques to fake the evidence:

    - The sock puppet technique is employed to mask the person's online identity by manipulation. By using several socket puppets, data related to a particular

person would be considered to have a position with various individuals. In this manner, the data collector would not have sufficient understanding to connect separate socket puppets to one person.
- Many security software could be used to disguise the identity of a person, for instance, Disguise Me. This is extremely helpful whenever the data owner has to send credit card info by online purchases.

3.4 Privacy of Big Data in the Data Storage Phase

In recent years, the data processing technologies have advanced, which include the growth of cloud computing, high volume in data storage is not really a major issue [18]. If a large data management device is breached, it may be extremely damaging as confidential information can be leaked [19]. In a distributed environment, an application can involve multiple datasets from multiple data centers, posing a privacy challenge. There are four types of traditional protection frameworks for protecting data. That includes data defense mechanisms at the archive level, data protection systems at the file level, security schemes at the media level, and encryption schemes at the device level [20]. The storage system needs to be flexible, in addition to the 3 V nature of big data analytics. It has the potential to be configured dynamically to accommodate a number of applications. Virtualization of storage, powered by the latest cloud computing paradigm [21], is one groundbreaking technique to deal with these problems. Virtualization of storage is a strategy that combines multiple network storage devices that provides the illusion of a single storage unit. SecCloud is a cloud data security model, both collectively known to be data storage privacy and auditing of cloud computing privacy [22].

3.4.1 Ways of Preservation of Privacy in Cloud Storage

As data is stored in the cloud, there are three main aspects of data security: confidentiality, trustworthiness, and availability [23]. For the first and the second, apply specifically to data privacy, like whether the confidentiality or dignity of the data is compromised, it would have a significant impact on the privacy of individuals. Access to information means ensuring that approved people are able to use the data as necessary. The primary prerequisite for large storage of data, the facility which comes down to safeguard the privacy of the user. There exist several processes in place to satisfy this obligation. Techniques to safeguard the privacy of consumers which are stored in the cloud are just as described:

- *Encryption dependent attribute*: Access management is the identification and is based on a person and directly available to all resources.
- *Homomorphic cryptography*: The IBE or ABE configuration settings can be used to configure the text cipher receiver.
- *Encryption of storage path*: Secures the availability of huge cloud data.

- *Hybrid cloud*: It is just a cloud storage platform that uses a combination of on-site, private cloud, as well as third-party, cloud resources in public that are distributed between the two systems

Verification of Integrity in the Storage of Big Data

But when it comes to cloud computing, it is being utilized for massive storage of data, the data proprietor lacks command of the data. The owner of the data has to be absolutely confident that the cloud stores the data rightly in keeping with the level of the contract of operation. To safeguard the confidentiality of the cloud user, the system must have a mechanism to allow the owner of the data to verify that his cloud-based data are preserved [24, 25]. In conventional schemes, the security of data storage can be checked by various methods like Reed-Solomon code, checksums, trapdoor hash functions, MAC, and digital signatures, etc. Verification of the precision of data is also very important. This contrasts with the multiple honesty confirmation systems presented [26].

3.5 *Privacy of Big Data in Processing of Data*

Model for big data analysis classifies batch, stream, graph, and machine learning systems [27, 28]. With respect to the data processing section's protection of privacy, the separation can take place in two steps. The purpose of the first step is to protect data from unrequested dissemination since confidential data from the data owner is contained in the data collected. Secondly, without breaching privacy, the purpose is to extract valuable information from the records.

3.5.1 Traditional Big Data Privacy Security Approaches

Few standard approaches for the defense of privacy in big data are listed briefly here. Predominantly, such methods get a certain degree of confidentiality, but the disadvantages nevertheless led to the emergence of new technologies.

De-Identification [29, 30]

It is a conventional privacy method preservation of the mining of data, where data must first be sanitized with generalization (substituting quasi-identifiers with lower precise yet semantically consistent variables) and suppression (not revealing any figures) prior to data release for mining. Mitigate risks from re-identification, conceptions of k-anonymity, l-diversity [31, 32], and t-closeness [33] are implemented for improving conventional data-mining protection of privacy. De-identification is an important method for the security of privacy and can be applied to the protection of privacy in large data analytics. However, an intruder will be able to gain many additional information aids due to de-identification but need to be conscious that big data will however give rise to the risk of re-identification. As the result, de-identification is not enough to secure the privacy of big data.

- Privacy-preserving in the mining of big data is also concerning, maybe due to the challenges of accessibility and efficiency or because of the threats of de-identification.
- De-identification seems to be more achievable for privacy-preserving big data mining while designing powerful privacy-preserving algorithms that aid minimize the probability of re-identification.

There are three mechanisms for the preservation of the privacy of de-identification, including K-anonymity, L-diversity, and T-closeness [34].

Comparative Study of Privacy De-Identification Approaches

Advanced data analysis can derive useful information with big data yet simultaneously there is a serious risk to the privacy of the user. Various techniques to protect privacy have been proposed prior, in the course, and after the big data analysis process. This paper addresses three mechanisms of privacy, like K-anonymity, L-diversity, and T-closeness. As the user's data expands fast and innovations keep on improving gradually, the trade-off between infringement and protection of privacy would be serious. Table 1 describes the current de-identification that preserves measures of privacy and the shortcomings in big data.

HybrEx

The model of hybrid execution [37] is a standard in cloud computing for confidentiality and privacy. It implements only operations via public clouds that are secure when incorporating the private cloud of an organization, i.e., public clouds are used for not sensitive data only and computing of an organization to be listed as public, while with regard to the sensitive, private, data and computing of an organization. The model uses its own private cloud. It ensures the accuracy of the data before the job is performed. It offers security integration. The following are the four sections in which HybrEx MapReduce provides for various types of software that use both public and private clouds.

1. *Map hybrid*: The mapping process is done in both public and private clouds, while the diminished process is carried out in one of the only clouds.
2. *Vertical partitioning*: Map and reduce functions are carried out using input as public data in the public cloud, swap intermediary data, and store the performance in the public cloud between them. A similar analysis is performed on a commercial server with private info. Jobs are being treated in solitude.

Table 1 Traditional privacy security mechanisms and their weaknesses in big data along with their computational complexity

Privacy practices	Descriptions	Constraints	Computing complexity
T-closeness	When the difference between the distribution of the essential attribute in that class and the distribution of the attribute in the whole table is not much greater than threshold t, the equivalence class is considered to achieve a T-closeness. If all of the equivalence classes have t-closeness, a table is considered to have t-closeness	T-closeness demands toward the distribution of the sensitive attribute in each equivalence class are identical to the distribution of the sensitive attribute in the overall table	$2^{O(n)O(m)}$ [35]
L-diversity	If the essential variable has a minimum of 'well-represented' values, the equivalence class is known to have L-diversity. If each table equivalence class has L-diversity, a table is considered to have L-diversity	L-diversity can be challenging and needless to obtain, and L-diversity is inadequate to prohibit disclosure of attributes	$O((n^2)/k)$
K-anonymity	It is a mechanism for designing and testing algorithms and frameworks that leak information in a way that the information revealed restricts what could be disclosed regarding the properties of entities to be secured	Background knowledge, homogeneity-attack	O(klogk) [36]

3. *Horizontal partitioning*: The Map Process is run on public clouds and only when the Reduced Process is run on a private cloud.
4. *Hybrid*: The Map Process and the Reduced Process are tested on both public and private clouds. Information exchange between clouds is indeed available.

Integrity audit models for maximum integrity and fast integrity checks are also suggested. The HybridEx problem was that it does not handle often with a key that's created in both public and private clouds in the mapping process, which only interacts as an oppressor with the cloud.

Privacy-Protecting Aggregation

This is based on homomorphic encryption utilized in standard data collection techniques for event statistics [38]. In the context of a homomorphic public-key encryption algorithm, various sources are using the identical public key to encrypt the personal data in such a ciphertext [39]. Such Cypher texts can be consolidated and also the consolidated answer could be obtained using the analogous private key. Still, aggregation is limited to intention. Privacy-preserving processing, however, will preserve user privacy in the processes of big data processing and storage. Owing to its rigidity, sophisticated data mining can not be used to leverage new insights. As just that, the privacy-preserving aggregation is insufficient for a broad data analysis. Operations on encrypted data: Motivated by the hunt for encrypted data, operations can be carried out over encrypted data to preserve the privacy of consumers of big data analytics. Since encrypted data mining is often time consuming and a great deal of time and large data are large-volume and enable us toward mining new information within a specific timeframe, the output of encrypted data operations can be described as incompetent in the analysis of big data.

4 New Privacy Security Approaches in Big Data

4.1 Differential Privacy

It is engineering that enables database analysts with a possibility to learn valuable information from databases containing confidential data about individuals apart from exposing the sensitive personal identities [40]. That is achieved by adding a minimal insights distraction transmitted through the database system. The implementation of distraction is broad enough in order to maintain secrecy and, simultaneously, quite tiny to ensure that the information given to the analyst remains relevant. Previously, several methods were used to preserve privacy, but have proven ineffective. Differential privacy (DP) provides a resolution to this problem, as shown in Fig. 3. Access directly to confidential records through direct access from the database is not given

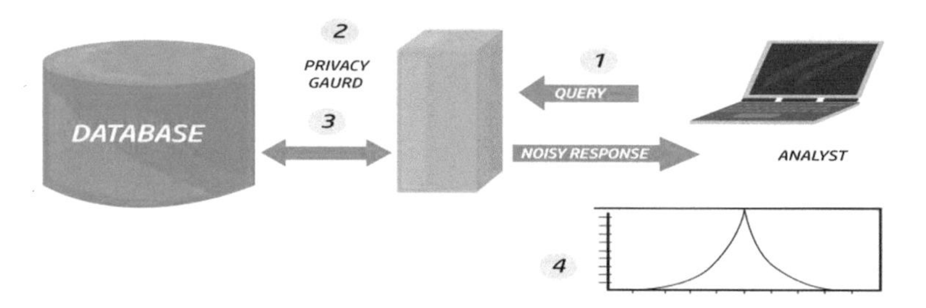

Fig. 3 Big data differential privacy (DP) as a private security approach for big data

to DP analysts. Between the database and the privacy analyst, an intermediate piece of software is introduced. This intermediate program is known as the privacy guard.

Step 1: Query is made by the analyst to the database via intermediary privacy guard.

Step 2: Query is taken by the privacy guard from the analyst and the privacy guard determines the privacy risk of this inquiry as well as other queries previous inquiries. After the privacy risk analysis.

Step 3: The answer is fetched from the database by a privacy guard.

Step 4: Some distortion is introduced to it on the grounds of the privacy risk calculated and eventually given to the analyst.

The extent to which privacy is assessed is proportional to the degree of distortion applied to the original data. If the risk to privacy is small, the additional prejudice is limited enough not to affect the report's accuracy but significant enough to preserve the user privacy of the database. Although where there is a significant risk of anonymity, then more distortion is applied.

4.2 Identity-Based Anonymization

Such approaches have faced difficulties by effectively integrating anonymization, preservation of privacy, and massive data techniques to evaluate usage data while preserving user identity. Intel has developed an anonymization architecture [41] that has authorized a range of tools to be utilized to de-identify as well as re-identify weblogs. This principle has demonstrated that large data strategies can provide benefits in the business world, particularly when operating on anonymized data. It looked at the benefits and drawbacks to fix such flaws and discovered that the User-Agent (Browser / OS) knowledge is closely associated with a single user. This is a case study on the application of anonymization in an organization, explaining the specifications, application, and knowledge faced by using anonymization to preserve privacy of business data analyzed using big data techniques. This standard anonymization research utilized k-anonymity based metrics. Intel utilized hadoop to analyze anonymized data and obtain valuable knowledge for researchers of the human factor. At the same time, they realized that anonymization had to be more than just masking or generalizing anonymized databases in those fields to carefully figure out if they were prone to assault [42, 43].

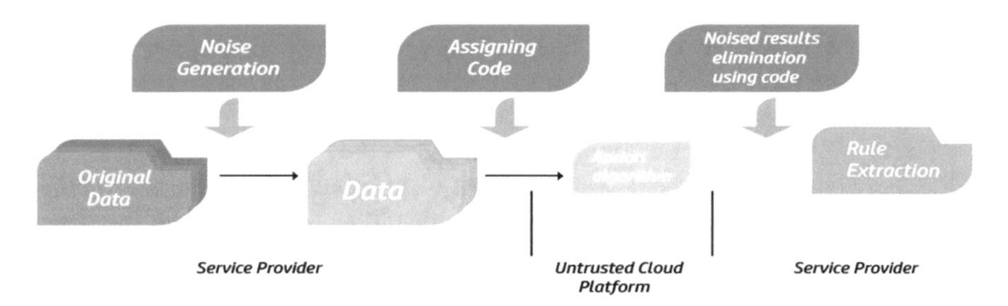

Fig. 4 Summarized association law mining method

4.3 Apriori Algorithm Privacy-Preservation Algorithm Within the MapReduce Framework

Hiding the needle inside a haystack Accepted privacy-preserving association rule algorithms change the initial transaction data by adding distortion. Although, such work preserved the initial transaction in the Noised Transaction in the taking note of the idea that perhaps the purpose is to do so avoid the erosion of the data value and at the same time avoiding a loss of privacy. The probability, however, of an unauthenticated cloud service provider inferring the actual frequent item collection persists throughout the method. Given the potential of association rule disclosure, include ample protection of privacy since this algorithm for the protection of privacy is built on the principle of "hide a needle in a haystack" [44]. Current techniques [45] can not hastily incorporate noise due to the need to recognize privacy-data value offsets. Alternatively, this approach incurs extra computing costs by incorporating noise that would make the "haystack" cover the "needle." Thus, making an arbitrage between issues will be easier to overcome by using the hadoop system in a cloud environment [46].

In Fig. 4, the service provider refers to the data collected in the original transaction by the data provider as a disturbance, a dummy entity. Consequently, a special code is given to the dummy and the original objects. To flush out the dummy object, the service provider keeps the code specifics until the standard object delivered by the external cloud platform is withdrawn. The Apriori algorithm is applied by a framework focused on the cloud that uses data generated by the service provider. Via the external cloud repository, a regular fixed asset and support package are returned to the service provider. It uses an algorithm to pick the frequent item selection to obtain the relevant partnership rule for the frequent itemset without the dummy item.

4.4 Privacy-Preserving Publishing Big Data

Essential components of corporate, educational, and clinical reporting, due to the increase in abundance of open sources available, such as accessing data from social networks and mobile devices, are the reporting and dissemination of real data, and over time the volume of this data has also increased [47]. Privacy-preserving models usually break into a couple of environments, which are known as input and output for privacy. The release of anonymized data for models such as k-anonymity and l-diversity is the primary issue of input security. There is a general interest in information security problems such as association rule hiding and query auditing where the efficiency of different data-mining algorithms is significant. Much of the privacy initiatives have centered primarily on the accuracy of privacy protection (vulnerability quantification), including the use of user data. Clearly, the option is to break the data into smaller sections (fragments) and independently anonymize each part [48].

Despite the fact that k-anonymity will discourage identity attacks, due to the absence of diversity of the related attribute inside the equivalence class, it does not protect against visibility attacks. The L-diversity model enables at least one well-formed answer value for each equivalence class. It is popular for massive data sets to be analyzed using distributed frameworks such as the MapReduce framework [49, 50] so as to spread an expensive operation across several nodes and gain a substantial increase in performance. Enhancements to privacy models were also made in order to fix inefficiency.

4.5 Improving the k-anonymity and I-diversity Privacy Model

Anonymization based on MapReduce: The MapReduce paradigm is introduced for effective data processing. Bigger data sets are managed for broad and disbursed MapReduce systems. The data is separated into equally sized parts, which would then be loaded into different mappers. Mappers process their components and give sets of output. Pairs of the same key are pushed by the machine toward a single reducer. Reducer output sets are then used for the final test to be obtained.

K-anonymity by MapReduce: The k-anonymization algorithm should be sensitive to the propagation of data by mappers, even if the data can be segregated on their own by the MapReduce method. The Mondrian algorithm is evocative of our MapReduce-based algorithm. Per equivalence class is divided into (at most) q equivalence classes in each iteration, instead of two, every equivalence classes in each iteration, in order to maximize the prevalence and, most importantly, decrease the required cycles [51].

L-diversity MapReduce-based: To expand the paradigm of privacy from k-anonymity to l-diversity, which often requires the integration of essential values into output keys or the values of the mapper. Therefore, it is important to upgrade pairs generated by mappers and combiners correctly. The mapper in l-diversity gets both the quasi-identifier and, as info, the sensitive attribute, compared to the mapper in k-anonymity [52].

5 Global Data Privacy Initiatives

With complexity in the way of data usage and an increase in the cases of a data breach, there is an ardent need for robust data protection laws. In today's era, the data can be moved in a cross-border global environment. Businesses are required to monitor the universally compliant legislations in a 360-degree view manner. Many countries have started incorporating regional data protection laws but there has been no such robust existence of a global norm for data protection [53].

GDPR: It is otherwise known as the General Data Protection Regulation. It is pan-European legislation that was passed by the European Parliament in 2016 and came into existence on May 25, 2018. It ensures the liberty of movement of user data throughout the European Union along with the protection of user rights to individual data. GDPR is not only for the businesses that operate primarily in the EU region but also for those who are operating outside the EU region if they use and process personal data of EU subjects and the activities for processing relates to (a) Paid or free offerings of services and goods for EU data subjects, or (b) monitoring the behavior of EU subjects within EU [54].

APEC: Asia pacific economic cooperation has twenty member economies some of which are Canada, Japan, South Korea, China, USA that together equate to 55% of the world's GDP. Several initiatives have been taken by the APEC to ensure data protection. APEC's Cross-border Privacy Rules Framework (CBPR) is one of APEC's main initiatives. It offers privacy management principles that can be used by companies in correlation with the private system of APEC. The primary purpose of this scheme is to promote cross-border data sharing by using a mechanism that guarantees minimal protection of privacy and security [55].

6 Access, Confirmation, and Correction Rights

Data protection laws should incorporate the right to access, correct, and confirmation of data. However, the implementation of these data protection and privacy principle rights particularly for their expense and implementation has now been recognized. Hence, a reasonable fee may be charged by the companies and organizations as determined by the DPA [56].

The right to confirmation involves the right to inquire regarding the handling of individual data by a trustworthy of a data principle. The right of a data principle to gain admittance to personal stored data in a data fiduciary is ensured by the right to access. This enables a data principle to gain access to the copy of all his/her personal data held by an entity. The main aim and basis of these rights are to ensure that the data principle is able to verify the lawfulness of the data processing.

The enforcement of substantive data fiduciary obligations for a data principle is ensured by the rights to access and confirmation. These rights can only be enforced only when the data principle knows what personal data of his/her is held by a data fiduciary and how it is used. It is very important to keep in mind that the substantive obligations can be just mere platitudes without the right to confirmation and access. Thus, the rights to access and confirmation should be in the proper place under the law. The rationale of these rights guides their scope. According to the previous paragraph, these rights are thus the rights that help a data principle to know the extent and scope of his personal data held by a data fiduciary. Consequently, these rights empower a data principle with the proper actions to be taken against a data fiduciary in case of any kind of data breach. They also act as instruments to check the legality of a data fiduciary's processing of personal data.

6.1 Rights to Objection, Restriction, and Portability

These rights ensure different ways or approaches to ensuring the lawfulness of data processing by providing the data principles with the power to hold them accountable data fiduciary. It is an extension of the autonomy principle which is considered and believed to be a key plinth for securing a fair and free digital nation. However, autonomy can be curbed not only in the favor of interests competing but in the interest of more efficient and effective public good achievement of a free and fair digital economy as it is not absolute. Such a framework requires the need to not only be fair in principle but also in practice. This implies the most efficient methods to verify and ensure the lawfulness of data processing and for the consented purpose. Such methods may not be always feasible for an individual.

Therefore, the treatment of these rights in the group can be divided into three categories. First, the data processing objection right, direct marketing objection, and processing restriction do not accommodate within the lawful processing framework established by the previously mentioned framework. The right to object can be enforced by an individual in EU GDPR according to his/her particular situation, where the lawful processing of individual data for a public interest, exercising the official power and legitimate interest is ensured. These grounds are not mentioned in a similar or likewise manner in this framework as is mentioned in the EU GDPR. Further, it may be difficult to prove such an objection valid by providing appropriate specific grounds. It is kind of a very vague approach for providing un-enumerated grounds for ensuring lawful data processing.

Finally, to the limit that data processing is in accordance with the law or it is in the forwarding of a non-consensual data processing ground, the onus of the data can shift to any of these cases once allowed. For example, the processing must be done in legal supervision if the Aadhar data can be processed in accordance with the Aadhar Act formulated by the Unique Identification Authority of India so that the integrity of the Central Identities Repository can be maintained. If the circumstances of an individual change such that those circumstances can render future processing then the individual can have a remedy. It appears ill-conceived when a predominant personal interest is created without delineating.

Data fiduciaries can only engage in direct marketing of data by taking the consent of data principles, which is clearly mentioned in the reinforced standards of the data privacy protection standards. Therefore, if the data principle is not in favor of giving consent for direct marketing of his/her data, then the data fiduciaries should not reach the respective data principle on any mode of communication with any kind of marketing material. This provides the path for the need of another right for the objection of direct marketing. Although TRAI captures a lot many direct marketing activities that involve both text messages and calls, direct marketing activities through social media accounts and emails go unchecked. Therefore, a consent-based framework for direct marketing purposes would encourage regulatory actions in each sector. Finally, it might be unnecessary when it comes to giving a right for processing restriction given the right given to address the interim issues such as the inaccuracy of data. Data principles can always reach to courts or DPA for requests to stay on data processing, but the added benefit of this right exercised against a data fiduciary is not yet clear. Needless to say, that without this right, will not derogate the data principal's ability to take back consent in any way for processing of data and leading to unlawful processing. The right to object to automated decision making and to access the logic for it can be availed by the second group of rights. These rights-based on legitimate and logical rationale are derived by the EU to address the emerging challenges in AI and big data. They primarily aim at curbing harms caused by discrimination and prejudice in outputted data without evaluation of determinations in the view of a human. The point put by this right is to add a layer of human review that is not in accordance with the prejudice immune. This can affect the operational structure of an organization. Such changes can be helpful and necessary only if they are provided to the specific organizations in correlation with the nature of their data processing activity. This can be better implemented in a framework structure which requires certain data fiduciaries that aim at weeding out discrimination by making evaluative decisions through automated means. This privacy design constituent element should be proactively implemented by entities. DPA should periodically perform auditing and monitoring to verify for unlawfulness in the data processing. The individual of an agency is not denuded by such a model or framework. Individuals have the liberty to always reach out to DPA or go to courts just in case of breach of data fiduciary activities if discrimination ensued is due to lawful, discriminatory automated data processing. Therefore, in order to achieve the interests underlying these rights effectively, there is a need for an accountability model.

The right to portability of data plays a major part in making the digital economy go seamless. This right empowers the data principles to allow shifts of their individual data kept by a fiduciary for data in a structured and readable format. Thereby, it increases the data principles' power by giving them the right to control the flow of their personal data. Further, the freedom of movement of individual data allows their transfer from one data fiduciary to another. This in turn can lead to increasing consumer welfare by increasing competition among the data fiduciaries in the same respective industry. As the right may result in access to individual data generated during the use of profiles or goods created on a particular data principle or in the provision of services, it might lead to the reveal of trade secrets of a particular data fiduciary due to the access to such information. It is important to guarantee the right even in the case of providing the profiles or data without disclosing relevant secrets relating to data principles. However, if it is not feasible to provide a set of information without the disclosure of secrets of data principles, the request may be denied. The right to transmit or transfer the data from one fiduciary to another or among fiduciaries is constrained by technical feasibility. That is, data fiduciaries are not bound to provide data portability if it can be proved by them that the technical constraints existing currently imply the transmission or transfer of data as unfeasible. The standard relating to the technical feasibility of providing data transferability can be set through the DPA-developed codes to ensure that data fiduciaries do not use this clause to deny the right to data portability for the data principles. Further, fiduciaries are allowed to charge a reasonable to implement this right effectively along with the correlation to cost concerns. In this way, the data portability right can be effectively and efficiently engaged and implemented by the data fiduciaries and thereby maintain a balance with technical, business, and financial constraints.

6.2 The Right to Be Forgotten

It is an idea that aims at the memory limitations in a digital world. A memory having limited boundaries along with the aspects of forgetting and remembering are the essential human facets. The Internet at the current instance with its vast data reservoirs gives a picture of limitless memory. Therefore, there is a need for the ability to forget. Collective efforts for forgetting has often led to history being rewritten and thus has led to not a completely undesirable asset. However, the user's power to apply for forgetting is a kind of autonomy expression that can lead to better data protection. This becomes important when data flows are approved and initiated by the individuals who should be free and to whom others should be completely fair. But when designing such a right, it should be considered that collective goods and individual freedoms can be impacted. Removing public information takes the right to know for an individual by instilling the freedom of the press which published the story at first. Further, if every individual starts to enforce the right to forgetting for various personal data, the public information stereotype can be brought to question as such acts can lead to permanent deletion of information. The risk that attracts major concern is that the deletion of information can be from both the private as well as a public sphere. Therefore, in order to address these concerns, we have to make a distinction between permanent deletion from storage or disclosure restrictions (like de-indexing from the search results). This can never be passed as a separate individual practicing right altogether.

6.3 Cross-Border Transfer of Personal Data

It is indeed important to have economic output and feasibility, effective law enforcement should be ensured for any kind of cross-border data transfer framework. However, this must not check the international flow of data unjustifiably which is considered as very beneficial for promoting the digital environment of a nation. This is like the physical economy where the restrictions, as well as free movement of goods, operate alongside as a composite. The main matter of concern is to determine which data can be approved for cross-border transfer.

It is only partially accurate to describe the status quo as such as to why the freedom to transfer the restrictions and rules are considered to be the exception to freedom. The right to exchange personal data in the public realm works selectively in certain early-moving nations. For instance, by using the virtue of advancement in technology, the USA can create a completely open digital sphere and digital economy without causing any kind of hamper to its national interest. There is a need for local enforcement in correlation with the personal jurisdiction that the United States of America imposes on many tech companies and data reservoirs along with the huge amount of data stored within its boundaries. Thus, any form of support for the freedom of the personal data transfer in the digital sphere and digital economy should be cautiously reviewed.

In the EU, the operability of rules that are prescriptive is constrained by the cross-border data transfers permissibility defined by a limited set of circumstances. These include transferability to jurisdictions where the norms for the protection of data are considered adequate, transfers that depend on the different forms of rules and clauses relating to contractual agreement, or other scenarios where the need for data transfer is substantial or there is reduced risk of harm. The EU has granted adequacy certification to only 12 countries in the world, the privacy shield system restricts the exchange of data with the USA. Bilateral agreements made approved and accepted among Australia, the USA, and Canada for sharing of data like airline passenger data for efficient and effective enforcement of the law. Most data sharing to deemed non-adequate countries are done and ensured at the company level are involved and are authorized stakeholders in the contracts with binding to corporate rules or any such standard clauses [47].

There are two sets of standards issued by the European Commission, first for data transfer between data controllers and second for processors transfer outside the EEA/EU. However, there have been criticisms for contractual clauses for not being able to be implemented, due to the difficulty that the DPAs face while identifying non-compliance. Further, the contractual clauses on the grounds of their validation have been referred to the EU Court of Justice, thereby leading to the uncertainty about the future of these transfers. On the other hand, the framework for the data transfer is provided by the corporate standards and rules within multinational companies, with the rules needing the approval from not only the DPA appointed which is considered as the primary authority for this issue but also from the other EU countries DPAs from where data transfer outside the EU occurs [49].

The alternatives that might be considered regarding the feasibility of cross-border data transfers have been mentioned in the discussion above. To understand where the line needs to be drawn, the interest in regulating these transfers is essential to be kept in one's mind. The starting point should be the evaluation of data types to which law is to be applied. This can be considered as the universal bunch of data to which cross-border data transfer rules can be applied in correlation to personal and territorial interests. As the scope of this law, the cross-border flow of other data types won't be governed by this law. Needless to say, such activity can be regulated by distinguishable concerns and a suitable policy can be designed by the Government, but such analysis can be defined as a kind of beyond remit of this discussion area.

Personal data maintained in a country will always have the protection of that country's data protection regime. However, at least an adequate amount of protection level should be done for the personal data transferred abroad in the national interest. Given the seamless transferability nature of personal data, if such a restriction will not be imposed then the efficiency of the substantive protections that are provided by the law can be compromised. Therefore, the rules that ensure such data protection standards should be enforced.

The concern that needs to be addressed is how can such a protection level be affected. Our discussions about the limited adequacy nature of the EU framework suggest that determinations based on an analogous model should be looked at cautiously. Though it can prove to be providing a range of merits for the entities that need their personal data transferability, it can cause undue burden for the law enforcement by regulators. A new entity possessing such capacity on Day one is highly unlikely. Rather the alternative adequate mode of rules or clauses must be improved, especially molding them in a form that provides adequate data protection and has a better in terms of its enforcement.

6.4 Benefits of Enforcement

The matter of data being stored locally is connected to domestic law enforcement as well as the laws for the protection of data. The bodies providing law enforcement and intelligence agencies play a very complex and challenging role in this twenty-first century. They are responsible for tackling cybercrime, prevention cyber attacks, checking the growth of terrorism. Access to personal data while investigating ordinary forms of crime is also required. Further, effective DPA enforcement is also required when it comes to the data fiduciaries' obligations on them being pursuant to the data protection framework. Law enforcement bodies often are required to access data kept by data trustees in order to satisfy this mandate. Therefore, it becomes important for the law to acknowledge the easy and quick access to various information and data in order to ensure public and national safety and security. Therefore, if the data are stored locally, then it can provide a boost for the law enforcement bodies to have access to information that is essential to detect the crime and to collect enough potential and non-potential evidence for prosecution. It is also better for law enforcement agencies to provide access to information in their territory than to seek information from organizations keeping data overseas. However, in the future, it is advisable for the nations to collectively design a framework that encourages information sharing within the law enforcement bodies through harmonization. Reservations for this argument are being mentioned in several academic writings. There are three claims that can be made in this regard: first, there won't be any hamper for the domestic enforcement of law due to the data not being available because in order to provide access to the information many laws require cloud-based service providers; second, business growth can have a negative impact as it is economically not feasible for the companies to encourage local processing of data; third, the law might not be enforced because the data fiduciaries would know it is difficult for the enforcement bodies in such a scenario. US entities own eight among the top ten Web sites that are accessed by the users in India currently. Therefore, there is a greater chance that the enforcement bodies for the purpose of investigation will have to request these US entities to have access to the data. Although there is not a clear estimate of the total number of requests made by the investigation agencies in India to these entities, at least the data relating to 53,947 separate users held by mainly US tech companies have been successfully sought out by the UK government in 2014. Further, Indian Governmental agencies requested Google for the access of 3843 user data between January to June in the year 2017 of which only 54% of the total cases data were made available. Thus, access to 46% of data cases was refused by Google.

It is vague to assume that with the availability of local data processing and storage regulations, law enforcement will become much efficient. This is because, despite the data being stored locally, it might lead to a law question conflict if the concerned entity's registered country or any other country with which there is a connection for the entity, jurisdiction can still be asserted. However, if personal data that lies under the data privacy and protection laws are processed in a country then the probability of refusal from the foreign entity to access the data can be severely reduced. Further, even if there is a refusal for the access, the data stored in a physical location would play a key factor to have court jurisdiction in the concerned matter. Thus, domestic law enforcement can be increased by the data stored and processed in local facilities and it can be done through preserving at minimum one backup of all personal data within the jurisdiction.

6.5 Building an AI Ecosystem

The growth of AI is heavily dependent on processing and controlling huge volumes of data that could be effectively assessed through a big data environment that primarily stores data locally. This is because, currently, most of the personal data of users is being stored in abroad locations by tech giants like Facebook, Google, etc. Azmeh and Foster in their 2016 study mention the benefits that developing countries can avail from the local data storage of data principles. This is in particular first, greater spending in digital infrastructure and second, benefits of server localization for processing and storage of data for the development of digital technology and digital business through the participation of trained practitioners and enhanced connectivity. Incipiency of digital infrastructure and the digital industry is crucial for the development in the field of AI and other disruptive technologies, therefore highlighting the need for a policy for the storage and processing of data locally. This benefit can indeed be obtained by guaranteeing the local preservation of at minimum one copy of personal data. The additional, better benefit can be availed by the processing of sensitive and critical data within the county territory.

7 Brief Review of Indian Personal Data Protection Bill, 2018s

This bill must be seen as a beginning and not as the ending of a phase. This turns out indeed an excellent start, yet it is a far way from being flawless. As the world's biggest democracy, India should aspire to maintain global standards for human rights and civil liberties. This bill proposes desperately required changes like the restriction of intent, restriction of collection, restriction of storage, design protection, accountability, safety protections, and so on. That being said, it falls short of achieving all the current ones rather than defining a higher norm. In the bill passed new fields like data location, cross-border data sharing, violation reporting and right to erasure are regressive. Intercepting messages, monitoring also, direct messaging have been addressed in being the leaked Privacy Bill of 2011, nevertheless, main concerns were totally lacking in the current bill [57]. Public hearings on the bill should be held, and necessary amendments must be introduced to enforce a statute that respects citizens' confidentiality rights. The bill shall apply to the compilation by an Indian (State, resident alternatively company integrated in India) of personal data inside the territories of India in connexion with any undertaking where goods and services are sold to persons residing or to the surveillance of persons residing in India. It may establish two independent bodies: the Data Protection Authority of India (DPAI) and an independent Appeal Tribunal. The bill grants undue power to the central government. The Central Government has the right to give guidance to the DPAI. The bill distinguishes among others that determine what to do with the data versus that over that process the data of a normal citizen ('data principal'). The parameters of the trust data and the data processor shall provide the State (government) in its sphere of operation [55]. That being said, the Government has numerous exceptions and mechanisms to overcome certain of the conditions imposed on others under the law are set out below. A further distinction was being concerned in the bill regarding de-identification and anonymizing. De-identification underneath the bill is a pseudonymization underneath the EU GDPR. That is a method to delete individual identity from every other detail and substitute this with anyone special non-personal identifier. When there's not a single identifier added to data and that there's no form of re-identification, the data will be made anonymous. The DPA does have the right to lay down requirements where such data can be rendered anonymous [58]. For now, though, the only data protection applicable underneath the law in India occurs in the form of privacy for confidential personal information underneath Section 43A of the Information Technology Act, 2000 and also the rules laid down therein. Because this bill aims to build a modern and inclusive system for data protection, thus it intends to abolish some outdated protections from the legislation. The amendment is often included in the context of the 2005 Access to Information Act. It seeks to expand its coverage by refusing RTI questions mostly on the basis of information pertaining to personal data that is able to impact the data controller. Under the bill, the definition of sensitive personal information has now been modified from the existing sense of Rule 3 of the Information Technology Act (Reasonable Privacy Policies and Practices and Safe Personal Data), Regulations, 2011 [59]. New changes to sensitive personal information usually involve transgender identity, intersex status, official identity, religion, sex life, caste or tribal status, certain political beliefs, and membership, as well as any sort of information identified by the authority. In order for the approval to be deemed legitimate, the bill demands that the approval must be free, aware, precise, explicit, and worthy of being revoked. The bill suggests that approval would be notified when different minute specifics were given to the data concept in a document.

(a) The reason for the collection of the data
(b) The period of time of data storage
(c) The processor with which the information is exchanged
(d) The freedoms to view and track the data
(e) Procedure for dispute settlement.

Extra information, for instance, the specifics needed by the aforementioned existing clause, may be provided access by reference at the end of a notice which says ‘More information’ either certain terms to that effect. That provision for just a comprehensive warning does not address its existing shortage of substantive and proper consent. The Data Principles were given the freedom to revoke permission, but “both practical implications for the results of any discontinuation” shall be met by the Data Principle itself. Quite apart from the foregoing, the idea of the agreement has also been substantially weakened throughout the Act. A broad exception was made for the collection of confidential and confidential personal information required for just about any activity of the Parliament or any State Legislature or any action of the State allowed underneath the legislation for the provision of every sort of credential or benefit. A specific exemption occurs for the collection of personal data in order to grant a registration, warrant, or permit. The collection of confidential and confidential personal information may also be carried out for “other role of the Parliamentary or any State Legislature” and “the practice by the legislation of any functionary throughout the delivery of any service and gain to just the Data Concept". The bill may give DPAI the right to usage of personal information for such reasons without permission, involving: avoidance and monitoring of illegal activity, such as bribery, whitewashing, mergers and acquisitions, network, and data protection, credit ratings, loan collection, and analysis of access to the public data [53]. Financial data, passwords, health, biometric and genetic data could be collected without permission throughout a breaking down in public order. The Data Values are given some rights in this bill in accordance with the rights provided to data subjects in the EU General Data Privacy Legislation. In addition to the certification of past or continuing data processing operations, the overview of the personal data being stored, data trustees may charge the data principals. The right to access data is extremely restricted. Rather than allowing the data manager and data processor to include a full data copy kept by the organization, the bill would allow them to only include an overview of personal information and a description of the production activity. In-kind of a different right to portability of data, the access to data produced by the data principal to the data trustee and to the data generated by the data trustee on the data principal has been made available. That being said, it is also significantly restricted except that the right to data portability could not be invoked so as to acquire a person's data once it has been stored due to another reason. Unwarranted exemption occurs when complying with the order for data portability "reveals the trading confidentiality of some data trustee or is not theoretically feasible." It leaves a vacuum for data trustees to not incorporate a framework for data portability across their goods and services, and then argues that perhaps the requirement for data portability could not be agreed with on the grounds of technological infeasibility. Intent restriction and compilation constraint are obsolete in the absence of a clear right to access one's records. Right to corrections is now only applicable "where needed." The definition also isn't specified in the bill. Data trustees may deny a request for corrections of the data if they decide that perhaps the correction is not required. Right to corrections is now only applicable "where needed." The definition also isn't specified in the bill . Data trustees may deny a request for corrections of the data if they decide that perhaps the correction is not required. We see no justification why trustees should be allowed to deny the appeal toward corrections of personal information even by data principals. The bill doesn't really allow enough freedom to consider removing and the freedom to be deleted only offers minimal relaxation to data principals who wish to delete/delete their data. Any specific protections against the manipulation of their data are being given to children, including arrangements for age verification and parental consent. Data trustees who promote children have not been authorized to carry out such data processing operations.

DPAI does have the authority to nominate such data trustees or groups or data trustees as essential data trustees. DPAI does have the authority to mandate major data trusts to meet any of these criteria: data checks, data security agents, maintenance of data, and data privacy risk assessments. The Act notes that perhaps the examiner can award a ranking to something like the data trustee throughout the form of a data confidence score. Data trustees will view this data confidence score.

Aadhaar has a connexion within the bill to the concept of an official identifier. Although, the use of this phrase is entirely needless since the expansive meaning to such an official descriptor throughout the bill prevents the provision that the word is specifically used in the framework of the law. The Unique Identification Authority of India includes no reference throughout the law. This does not really imply that UIDAI and Aadhaar are not affected by the bill. Data collection practices conducted by

UIDAI are being given validity under the Statute, although the publication of data has been prohibited without clear reference to Aadhaar. The Central Government has the authority to give instructions to DPAI "for the sake of the dignity and independence of India, the protection of the State, good ties with foreign states or public order" and "instructions on policy issues." Any operations by data processors in India can be exempted by the Central Government if they contribute to data of data principals located outside India under a contract with data trustees. The Central Government may ban the procurement of any biometric data unless that processing is allowed by statute. Transborder data sharing is permitted under some provisions, however, a backup of the personal information should be maintained in India. In fact, the central government (not the DPAI) may inform those types of personal data that are processed only in India. Transborder movements of data may be granted pursuant to normal contractual provisions, intragroup agreements, or to countries authorized by the Central Government in conjunction with the DPAI. Consent is essential for the transmission of data across boundaries. The data localization provisions of the bill do not serve to protect the privacy of consumers in any way whatsoever. Data localization would impact smaller players in business which may be a big concern for startups in India if other jurisdictions still follow suit. Many privileges and responsibilities in this bill are unenforceable if data are collected for Public security purposes; avoidance, identification, review, and prosecution of contraventions of law, domestic purposes, journalistic purposes, or legal proceedings.

Data infringement alerts shall be rendered to DPAI by data trustees if the violation is prone to injury to any data principal. The notice of the infringement shall contain the existence of the personal data, the amount of data principals involved, the potential effects of the infringement, and the action were taken by the data trustee to resolve the infringement. Subsequently, DPAI will decide if the data principles ought to be told of the data violation. The data trustees are not obligated to warn the data principals of the violation until they are ordered through DPAI. Data stewards do not require to warn the principals of the infringement, though the violation can induce significant damage to the data principal, without it being imminent. Data guardians will not need to warn the data principals of the violation, even though the violation is likely to cause serious injury to the reporting entity until the data principal takes urgent disciplinary action. DPAI has the authority to control and implement the bill, issue codes of practice and guidelines, perform inspections, issue notices, modify the authorization of enterprise or operation, and enact penalties. There will be three phases of the decision-making process underneath the bill. The first step was its adjudication by an adjudicator. Regrettably, the bill understands the necessity to separate the adjudicative DPAI wing compared with the other part of authority, but not from the central government themselves. The following stage in the adjudication process involves an appeal from the decision of the Adjudicator to the Impartial Judicial Tribunal which would have been formed under the bill. It is left to the powers, appointments, terms of office, allowances and compensation, resignation, discharge, and some other terms of service of the Chairperson and the other members of the Appellate Tribunal. Therefore, the second level of the adjudication process is not really exempt from discrimination. This renders the third and final level of the adjudication process entirely unbiased: appeals from the Appellate Tribunal can be placed before the Supreme Court of India. Penalties range from INR 5000/-per day of default to INR 15,000.000/-or 4 percent of global revenue, either of them is greater. Any principal data that has suffered damages due to the breaking of a law by the data processor or data trustee can be remunerated. Specific offenses underneath the bill are punishable by imprisonment for up to 5 years. Both violations of the bill are recognizable and non-compensable. The criminal provisions of the bill, many of which are not enforceable, could theoretically lead to prosecutions under Section 66A of the IT Act. Judicial requirements may be restricted to cases of sale/offer for sale of confidential/sensitive personal information.

8 Conclusion

Big data [2, 46] is evaluated for every bit of insight which contributes in order to provide improved decision making and pragmatic actions with regard to overwhelming enterprises. However, just a significant amount of samples is currently processed. Throughout the paper, we looked at the privacy issues in big data through defining its privacy standards along with subsequently addressing the current privacy-preserving strategies appropriate for big data. The difficulties of privacy in and step of the big data lifecycle were raised together by means of said benefits and downside of current privacy-conserving solutions throughout big data implementations. The present paper further introduces both conventional and modern strategies for the safeguard of privacy in big data. For privacy protection using association rule mining, the principle of hiding a needle amongst haystack was discussed. Ideas of identity-based anonymization and differential privacy and a retrospective analysis of different recent strategies of big data privacy was also explored. It provides robust anonymization methods in the MapReduce context, Which could quickly be expanded by expanding the number of mappers and reducers. In the future path, insights are required to achieve successful solutions to both the robustness of privacy and protection issues throughout the age of big data and, in particular, the task of conciliating security and privacy models by leveraging the map reduces the scope. Then, in this paper, policies and rights regarding big data standards such as Cross-Border Transfer of Personal Data, The Right to be Forgotten, Rights to Objection, Restriction, and Portability were also discussed. Along with that, a brief review of the Indian Personal Data Protection Bill of 2018 was presented. Differential privacy is one of the spheres that have a great deal of latent ability to be further manipulated. There are a lot of issues when the Internet of things and big data came along, even with the exponential development of IoT; the data volume is high, but the accuracy is poor and information belongs to multiple data vendors, who undoubtedly have a wide range of different styles and modes of representation. And data, as structured, semi-structured, and sometimes entirely unstructured, is heterogeneous. This raises new threats to privacy and brings up analysis concerns. Thus, going forward, it is possible to research and implement new ways of protecting mining safety. As such, there exists a great deal of space for more study into privacy conservation approaches in big data.

References

1. Kolomvatsos K, Anagnostopoulos C, Hadjiefthymiades IS (2015) An efficient time optimized scheme for progressive analytics in big data. Big Data Res 2(4):155–165
2. Abadi DJ, Carney D, Cetintemel U, Cherniack M, Convey C, Lee S, Stonebraker M, Tatbul N, Zdonik SB (2003) Aurora: a new model and architecture for data stream management. VLDB J 12(2):120–139
3. Big data at the speed of business [online]. https://www-01.ibm.com/soft-ware/data/bigdata/ 2012
4. Manyika J, Chui M, Brown B, Bughin J, Dobbs R, Roxburgh C, Byers A (2011) Big data: the next frontier for innovation, competition, and productivity. Mickensy Global Institute, New York, pp 1–137
5. Gantz J, Reinsel D (2011) Extracting value from chaos. In: Proc on IDC IView, p 1–12
6. Tsai C-W, Lai C-F, Chao H-C, Vasilakos AV (2015) Big data analytics: a survey. J Big Data Springer Open J
7. Mehmood A, Natgunanathan I, Xiang Y, Hua G, Guo S (2016) Protection of big data privacy. In: IEEE translations and content mining are permitted for academic research
8. Jain P, Pathak N, Tapashetti P, Umesh AS (2013) Privacy preserving processing of data decision tree based on sample selection and singular value decomposition. In: 39th international conference on information assurance and security (IAS)
9. Qin Y et al (2016) When things matter: a survey on data-centric internet of things. J Netw Comp Appl 64:137–153
10. Fong S, Wong R, Vasilakos AV (2016) Accelerated PSO swarm search feature selection for data stream mining big data. IEEE Trans Services Comput 9(1)
11. Middleton P, Kjeldsen P, Tully J (2013) Forecast: the internet of things, worldwide. Gartner, Stamford
12. Hu J, Vasilakos AV (2016) Energy big data analytics and security: challenges and opportunities. IEEE Trans Smart Grid 7(5):2423–2436
13. Porambage P et al (2016) The quest for privacy in the internet of things. IEEE Cloud Comp 3(2):36–45
14. Jing Q et al (2014) Security of the internet of things: perspectives and challenges. Wirel Netw 20(8):2481–2501
15. Han J, Ishii M, Makino H (2013) A Hadoop performance model for multi-rack clusters. In: IEEE 5th international conference on computer science and information technology (CSIT), pp 265–274
16. Gudipati M, Rao S, Mohan ND, Gajja NK (2012) Big data: testing approach to overcome quality challenges. Data Eng 23–31
17. Xu L, Jiang C, Wang J, Yuan J, Ren Y (2014) Information security in big data: privacy and data mining. IEEE Access 2:1149–1176
18. Liu S (2011) Exploring the future of computing. IT Prof 15(1):2–3
19. Sokolova M, Matwin S (2015) Personal privacy protection in time of big data. Springer, Berlin
20. Cheng H, Rong C, Hwang K, Wang W, Li Y (2015) Secure big data storage and sharing scheme for cloud tenants. China Commun 12(6):106–115
21. Mell P, Grance T (2009) The NIST definition of cloud computing. Natl Inst Stand Technol 53(6):50

22. Wei L, Zhu H, Cao Z, Dong X, Jia W, Chen Y, Vasilakos AV (2014) Security and privacy for storage and computation in cloud computing. Inf Sci 258:371–386
23. Xiao Z, Xiao Y (2013) Security and privacy in cloud computing. IEEE Trans Commun Surv Tutorials 15(2):843–859
24. Wang C, Wang Q, Ren K, Lou W (2010) Privacy-preserving public auditing for data storage security in cloud computing. In: Proceedings of IEEE international conference on INFOCOM, pp 1–9
25. Liu C, Ranjan R, Zhang X, Yang C, Georgakopoulos D, Chen J (2013) Public auditing for big data storage in cloud computing—a survey. In: Proceedings of IEEE international conference on computational science and engineering, pp 1128–1135
26. Liu C, Chen J, Yang LT, Zhang X, Yang C, Ranjan R, Rao K (2014) Authorized public auditing of dynamic big data storage on the cloud with efficient verifiable fine-grained updates. In: IEEE trans on parallel and distributed systems, vol 25, no 9, pp 2234–2244
27. Xu K et al (2015) Privacy-preserving machine learning algorithms for big data systems. In: IEEE 35th international conference on distributed computing systems (ICDCS)
28. Zhang Y, Cao T, Li S, Tian X, Yuan L, Jia H, Vasilakos AV (2016) Parallel processing systems for big data: a survey. In: Proceedings of the IEEE
29. Li N et al (2007) t-Closeness: privacy beyond *k*-anonymity and *L*-diversity. In: IEEE 23rd International Conference on Data Engineering (ICDE)
30. Machanavajjhala A, Gehrke J, Kifer D, Venkitasubramaniam M (2006) L-diversity: privacy beyond *k-anonymity.* In: Proceedings 22nd international conference data engineering (ICDE), p 24
31. Ton A, Saravanan M Ericsson research [Online]. https://www.ericsson.com/research-blog/data-knowledge/big-data-privacy-preservation/2015
32. Samarati P (2001) Protecting respondent's privacy in microdata release. IEEE Trans Knowl Data Eng 13(6):1010–1027
33. Samarati P, Sweeney L (1998) Protecting privacy when disclosing information: k-anonymity and its enforcement through generalization and suppression. Technical Report SRI-CSL-98–04, SRI Computer Science Laboratory
34. Sweeney L (2002) K-anonymity: a model for protecting privacy. Int J Uncertain Fuzz 10(5):557–570
35. Meyerson A, Williams R (2004) On the complexity of optimal k-anonymity. In: Proceedings of the ACM symposium on principles of database systems
36. Bredereck R, Nichterlein A, Niedermeier R, Philip G (2011) The effect of homogeneity on the complexity of k-anonymity. In: FCT, pp 53–64
37. Ko SY, Jeon K, Morales R (2011) The HybrEx model for confidentiality and privacy in cloud computing. In: 3rd USENIX workshop on hot topics in cloud computing, HotCloud'11, Portland
38. Lu R, Zhu H, Liu X, Liu JK, Shao J (2014) Toward efficient and privacy-preserving computing in the big data era. IEEE Netw 28:46–50
39. Paillier P (1999) Public-key cryptosystems based on composite degree residuosity classes. In: EUROCRYPT, pp 223–238
40. Microsoft differential privacy for everyone [online] (2015). https://download.microsoft.com/…/Differential_Privacy_for_Everyone.pdf
41. Sedayao J, Bhardwaj R (2014) Making big data, privacy, and anonymization work together in the enterprise: experiences and issues. In: Big Data Congress
42. Yong Yu et al (2016) Cloud data integrity checking with an identity-based auditing mechanism from RSA. Future Gener Comp Syst 62:85–91
43. Oracle Big Data for the Enterprise (2012) [online]. https://www.oracle.com/ca-en/technoloq ies/biq-doto
44. Hadoop Tutorials (2012) https://developer.yahoo.com/hadoop/tutorial
45. Fair Scheduler Guide (2013). https://hadoop.apache.org/docs/r0.20.2/fair_scheduler.html
46. Jung K, Park S, Park S (2014) Hiding a needle in a haystack: privacy-preserving Apriori algorithm in MapReduce framework PSBD'14, Shanghai, pp 11–17

47. Ateniese G, Johns RB, Curtmola R, Herring J, Kissner L, Peterson Z, Song D (2007) Provable data possession at untrusted stores. In: Proceedings of international conference of ACM on the computer and communications security, pp 598–609
48. Verma A, Cherkasova L, Campbell RH (2011) Play it again, SimMR!. In: Proceedings IEEE Int'l conference cluster computing (Cluster'11)
49. Feng Z et al (2014) TRAC: truthful auction for location-aware collaborative sensing in mobile crowdsourcing INFOCOM. Piscataway, IEEE, pp 1231–1239
50. HessamZakerdah CC, Aggarwal KB (2015) Privacy-preserving big data publishing. ACM, La Jolla
51. Sweeney L (2002) k-anonymity: a model for protecting privacy. Int J Uncertain Fuzziness Knowl Based Syst 10(5):557–570
52. Wu X (2014) Data mining with big data. IEEE Trans Knowl Data Eng 26(1):97–107
53. Mishra S, Mallick PK, Jena L, Chae GS (2020) Optimization of skewed data using sampling-based preprocessing approach. Front Public Health 8:274. https://doi.org/10.3389/fpubh.2020.00274
54. Zhang X, Yang T, Liu C, Chen J (2014) A scalable two-phase top-down specialization approach for data anonymization using systems, in MapReduce on the cloud. IEEE Trans Parallel Distrib 25(2):363–373
55. Dutta A, Misra C, Barik RK, Mishra S (2021) Enhancing mist assisted cloud computing toward secure and scalable architecture for smart healthcare. In: Hura G, Singh A, Siong Hoe L (eds) Advances in communication and computational technology. Lecture Notes in Electrical Engineering, vol 668. Springer, Singapore. https://doi.org/10.1007/978-981-15-5341-7_116
56. Zhang X, Dou W, Pei J, Nepal S, Yang C, Liu C, Chen J (2015) Proximity-aware local-recoding anonymization with MapReduce for scalable big data privacy preservation in the cloud. IEEE Trans Comput 64(8)
57. Chen F et al (2015) Data mining for the internet of things: literature review and challenges. Int J Distrib Sens Netw 501:431047
58. Mohapatra SK, Nayak P, Mishra S, Bisoy SK (2019) Green computing: a step towards eco-friendly computing. In: Emerging trends and applications in cognitive computing, pp 124–149. IGI Global
59. Mallick PK, Mishra S, Chae GS (2020) Digital media news categorization using Bernoulli document model for web content convergence. Pers Ubiquit Comput. https://doi.org/10.1007/s00779-020-01461-9

Chapter 7
Privacy-Preserving Cryptographic Model for Big Data Analytics

Lambodar Jena, Rajanikanta Mohanty, and Mihir Narayan Mohanty

1 Introduction

Traditional data processing methods are not sufficient for big data [1, 2] that contains data sets which are very large and complex. It is the collection of large volume of structured, semi-structured, and unstructured data. Because of the recent advancements in technology, amount of data are gathered from various sources like Internet, healthcare data, social media networking, IOT, and sensor data. The volume of data is increasing day by day. It is well explained by veracity, velocity, volume of data in terms of audio, video, text, or image [3]. Big data cannot be processed and analyzed in a traditional manner, due to their complexity [4]. To describe big data, 5 V model is often used. There are many properties associated with big data. The prominent aspects are volume, variety, velocity, variability, and value.

L. Jena (✉)
Department of Computer Science and Engineering, Siksha 'O' Anusandhan (Deemed to be) University, Bhubaneswar, India

R. Mohanty
Department of Information Technology, PSIT, Kanpur, Uttar Pradesh, India

M. N. Mohanty
Department of Electronics and Communication Engineering, Siksha 'O' Anusandhan (Deemed to be) University, Bhubaneswar, India

P. K. Das et al. (eds.), *Privacy and Security Issues in Big Data*, Services and Business Process Reengineering, https://doi.org/10.1007/978-981-16-1007-3_7

Volume: Big data name is defining this characteristic itself which is related to size that crossed from peta bytes to zeta bytes. In general, volume refers to the huge amount.

VOLUME = Very Large Amount Of Data

Variety: It is the different types of data which are collected for better calculations. This could be structured data or unstructured data as well. These data are not easy to handle by traditional data analytics systems.

VARIETY = Produce Data in Different Formats

Velocity: New data need to be managed as well, so velocity defines the speed required to generate and process data under appropriate time. In today's era, this can be easily done in real time with new technologies.

VELOCITY = Produce Data at Very Fast Rate

Veracity: In simple word, it is the authenticity of the data. There will be no need to process those data for which you are not much confident that it will return some meaningful knowledge or not. It refers to the quality or correctness or accuracy of captured data. That data should give correct business value.

VERACITY = The Correctness Of Data

Value: Irrespective of how much data is available, it should must hold some meaningful value which can be useful for an organization; otherwise, it make no value. So the data must hold valuable information.

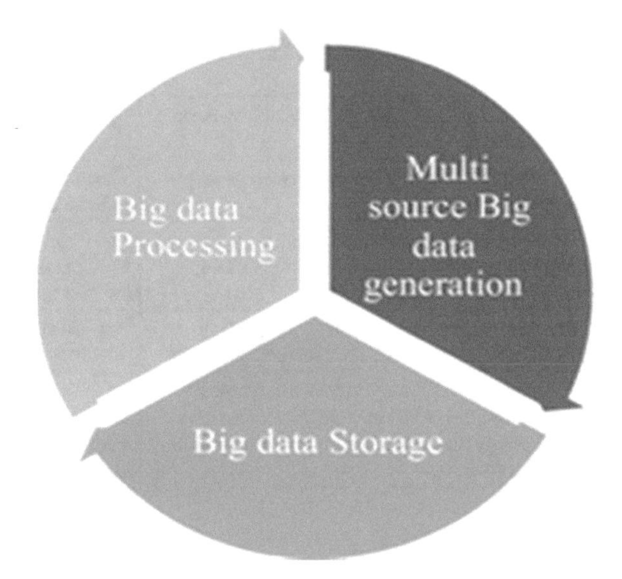

Fig. 1 Stages of big data life cycle

It needs the analysis of customer behaviour, market trend, and business trend of enterprises and organizations. Users can take decisions on their strategies due to the transparency in the functional big data analysis.

Various techniques have been developed in order to maintain privacy in big data. According to the different stages of big data life cycle [5], these techniques are grouped, namely data generation, data storage, and data processing. This is shown in Fig. 1.

According to [6], in data generation phase, privacy access protection and forge mechanisms are used. The encryption methods are well aligned to data storage phase which are used for privacy protection. Privacy preserving data publishing (PPDP) and data analytics from the big data are incorporated with the data processing phase. Suppression and generalization like anonymization techniques are used to protect the privacy of data.

1.1 Big-Data Privacy and Security

Privacy and security in big data in the event of complex applications are important issues. The information security of individual and personal data of an user is the serious issue during transmission over the Internet [7]. The security of data focuses on protecting the data from misuse and vulnerable attacks [8]. The difference has been presented in Table 1.

In big data, privacy and security are the major concern for user's data while they communicate with each other in a distributed environment. Confidentiality,

Table 1 Difference between privacy and security of data

Data security	Data privacy
Confidentiality, integrity, and availability of data is the security	Appropriate use of user's information is the privacy
The ability to be confident that decisions are respected refers to security	The ability to decide what information of an individual goes where refers to privacy
Providing security to the confidential data of enterprise is the main goal. Security system protects data of an agency [9]	The issue of privacy is one that often applies to a consumer's right to safeguard their information from any other parties
However, it is difficult to have good privacy practices without a good data security program	Possibility of poor privacy and good security practices
The company XYZ uses various techniques (Encryption, Firewall) in order to prevent data compromise from technology or vulnerabilities in the network	XYZ Company cannot sell user's shipping information (e.g., payment and address information) to a third party without prior permission of the user [7]

availability, and integrity are three dimensions toward data security stored in cloud [10].

While the big data snowball is speeding down the mountain of technical era to gain speed and volume, companies are trying to keep up with it. And they go downstairs, completely forgetting to put on masks, protective hats, gloves, and sometimes even skiing. Other than that, it is very easy to never cut it down by one piece. And putting all the precautionary measures at high speed can be too late or too difficult. Prioritizing low data security and putting everything up to the latest stages of big data acquisition projects could be a risky move. Big data security is defined by all the tools and technologies required to monitor any kind of attack, theft attempt, or other security breaches. Like every other cyber security attack, big data can be compromised from online or offline domains. These threats include the theft of individual data or an entire organization. There could be indirect attacks as well like DDoS attack which can crash the server. During big data analysis, the private information of individuals collected by social networks or feedback needs to be merged with huge data sets to find meaningful patterns, sometimes unintentionally in the whole process, confidential fact about a person might become open to the world. Often, it lead to privacy risk and violation of privacy rights. Some hackers or thieves who know better about big data take advantage of those who do not know much about this technology. Some big data technical issues and challenges are as follows:

- Processes need to be divided into smaller tasks and allocate these tasks to different node for computation purpose.
- Treat a node as a supervising node and check all other assigned nodes to see if they are functioning properly.
- Fault tolerance.
- Data heterogeneity
- Data quality
- Scalability

As big data is continuously growing in size day by day, so their concerns like security and privacy preservation are also rising regularly.

(a) The major reason for security and privacy concerns in big data is that now it is easily available and accessible to everyone. It has become so common that scientists, doctors, business executives, government employees, and ordinary people are also sharing data on a large scale every day. However, the tools and technologies developed to date to handle these vast volumes of data are not sufficient to provide better security and privacy to the data.
(b) Nowadays, available technologies are not sufficient to handle security and privacy threats; they lack the training as well as many adequate features and basic fundamentals to secure these vast amounts of data.
(c) Big data does not have much adequate policies that guarantee security and privacy measures.
(d) Technologies are not much capable of maintaining security and privacy, leading to many cases daily where they get tampered intentionally or accidentally. Thus, it is required to improve current algorithms and approaches to prevent data leakage.
(e) There is a lack of funding in the security sector by a company to protect their crucial data. It turns out that a company should spend at least 10% of its IT budget on its security but on average, less than 9% is being spent, making it harder for itself to protect its data.

Some important security and privacy concerns related to big data are as follows:

- Secure data storage and transaction logs.
- Security practices for non-relational data stores
- Secure computations in distributed programming frameworks
- End point input validation/ filtering
- Real-time security monitoring
- Scalable and composable privacy-preserving data mining and analytics
- Cryptographically enforced data-centric security
- Granular audits

2 Related Work

Many researchers have done various profound researches on security and privacy methods in big data. They proposed big data security and privacy in three categories such as input privacy, output privacy, and data security.

The anonnymized data of l-diversity and k-anonymity is the main concern in input privacy. A fast anonymization algorithm has been by Mohammadian [11] for big data security in order to minimize information loss. The randomization approach for privacy is referred in Evfimievski [12]. Each user keeps their personal and private data when they communicate the data to the server site. Only the properties of data which are important for statistical analysis are sent after randomization. It can be combined with the statistical approaches and cryptography. A profound work has been presented on social-network data in Tripathy [13]. Here, anonymization is achieved by mechanisms like l-diversity and k-anonymity. The k anonymity process is improved in Jain et al. [14]. The improved algorithm is implemented on a big candidate election data set. Somayajulu [15] works on data perturbation technique that can be used to rotate and shift the data values in attributes of a specific column. For map-reduce process, a multi-dimensional algorithm of k-anonymization is proposed by Zakerzadeh et al. [16]. In [17], an anonymization technique based on map-reduce is proposed.

Hiding of association rules and auditing of query are addressed in output privacy for privacy preservation. The Airavat model for privacy and security based on map-reduce framework is presented in Roy [18]. It integrates differential privacy (DP) and mandatory access control (MAC). Derbeko [19] has suggested a map-reduce method for hybrid clouds. Heuristic-based algorithm has been suggested in Chaudhari and Tiwari [20] in order to hide rules. A data hiding technique work in form of bits or pixels on a grid is presented in Yadav and Ojha [21].

Data security refers to the "confidentiality, integrity and availability" of data. In Terzi [22], there is a mechanism that provides the ability observe the secure communication over the network. It offers the authorized access to the systems to different users. Various security issues in cloud computing environment are described in Kacha [23, 24]. Most of the issues are raised due to single cloud, data life cycle, and different attributes such as "confidentiality, integrity and availability" [25]. Ilavarasi [26] addressed the security concern on distribution of micro data. The privacy-preserving data mining (PPDM) algorithms have achieved an increasing growth in order to provide privacy to personal data of any individual. The compromise between the data utility and privacy has been enhanced.

Knowledge, wisdom, and valuable information can be extracted from big data by advanced big data analytics. It is also equally important to provide privacy to user's data. Researchers have proposed numerous approaches for privacy preserving during the big data analytics process. The methods such as L-diversity, T-closeness, and K-anonymity are implemented by many authors [27, 28]. A detail study has been presented in Table 2 for big data privacy measures and the limitations.

Table 2 Data privacy measures and the limitations

Privacy measure	Limitations	Complexity
K-anonymity	Prone to attacks such as homogeneity attack, background knowledge attack	O(K ln K) [29]
L-diversity	Redundant and laborious to achieve, prone to attacks such as skewness attack and similarity attack	$O((n^2)/k)$
T-closeness	Hard to identify the closeness between knowledge gained and t-value	$2^{O(n)O(m)}$ [30, 31]

2.1 Hadoop Secure Map Reduce Model

In [32], the authors have proposed a hadoop secure map-reduce mechanism in order to fill the security and privacy gaps. The secure map reduce mechanism provides a way in securing the environment in distributed computing. Their model is depicted in Fig. 2 that introduces a privacy layer between hadoop distributed file system (HDFS) and map reduce layer of big data.

A lightweight encryption process is proposed for secure map reduce layer [33]. Here, the actual data is sent to hadoop distributed file system (HDFS) and then data goes to map reduce layer from HDFS. After entered into the map-reduce layer, encryption process begins.

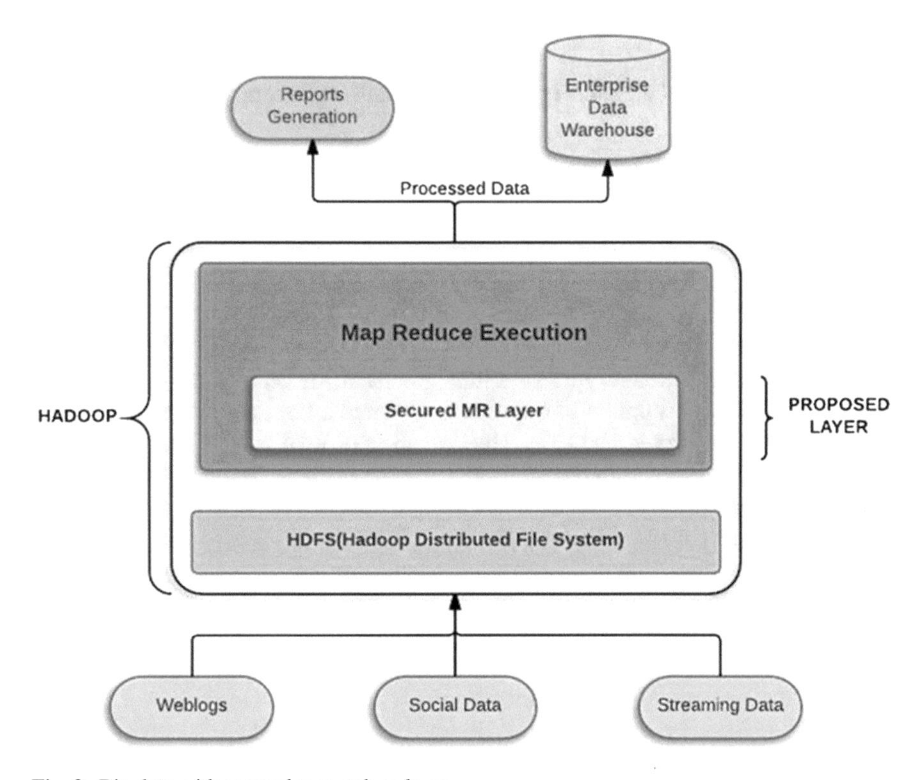

Fig. 2 Big data with secured map reduce layer

3 Proposed Cryptography-Based Environment

For achieving optimal privacy, cryptography-based algorithms are mixed together. The algorithms such as symmetric key and public key encryption and decryption techniques are proposed to achieve privacy in big data.

3.1 Data Encryption Standard

It is the most widely used block cipher. It maps a 64 bits of plain input to a 64 bits of encrypted output using 56 bits key [34]. It is base4d on two characteristics: transposition (diffusion) and substitution (confusion). There are 16 number of steps, each step refers to a round.

The details of each block of Fig. 3 are shown in the following permutation tables: Table 3a, b.

DOUBLE DES: It repeats DES twice using two keys K1 and K2 [35, 36]. The functionality is shown in Figs. 4 and Fig. 5.

3.2 RSA Algorithm

To encrypt the text in RSA crypto algorithm, public key is generated by using prime numbers. The key is supplied to the user to generate a key for decription in order to decypher the encrypted text [37]. The limitation of RSA is that easily the prime numbers can be guessed with few trials in brute-force attack. Thus, further developments are required to generate key using complex mechanisms which will increase security and provide more privacy [38].

3.3 Encryption and Decryption at Multiple Levels

Most of the existing systems aim at single-level encryption but the proposed system has been developed to support multi-level encryption techniques. Here, a function is used for generating random number of n digits. It is developed in such a way that if n numbers of algorithms are used, then it will generate an n digit random number. This number consists of the digits from 1 to n. From the generated random number, it will be decided in which order the algorithms are to be implemented for encryption. Since the number is a random number, it is not feasible to decode the order of execution by the eavesdroppers. Thus, the proposed technique provides stronger security to the sensitive data. The intruder faces the worst scene to crack the encryption order. The crypto-systems type, number, and order of execution will remain very hard to

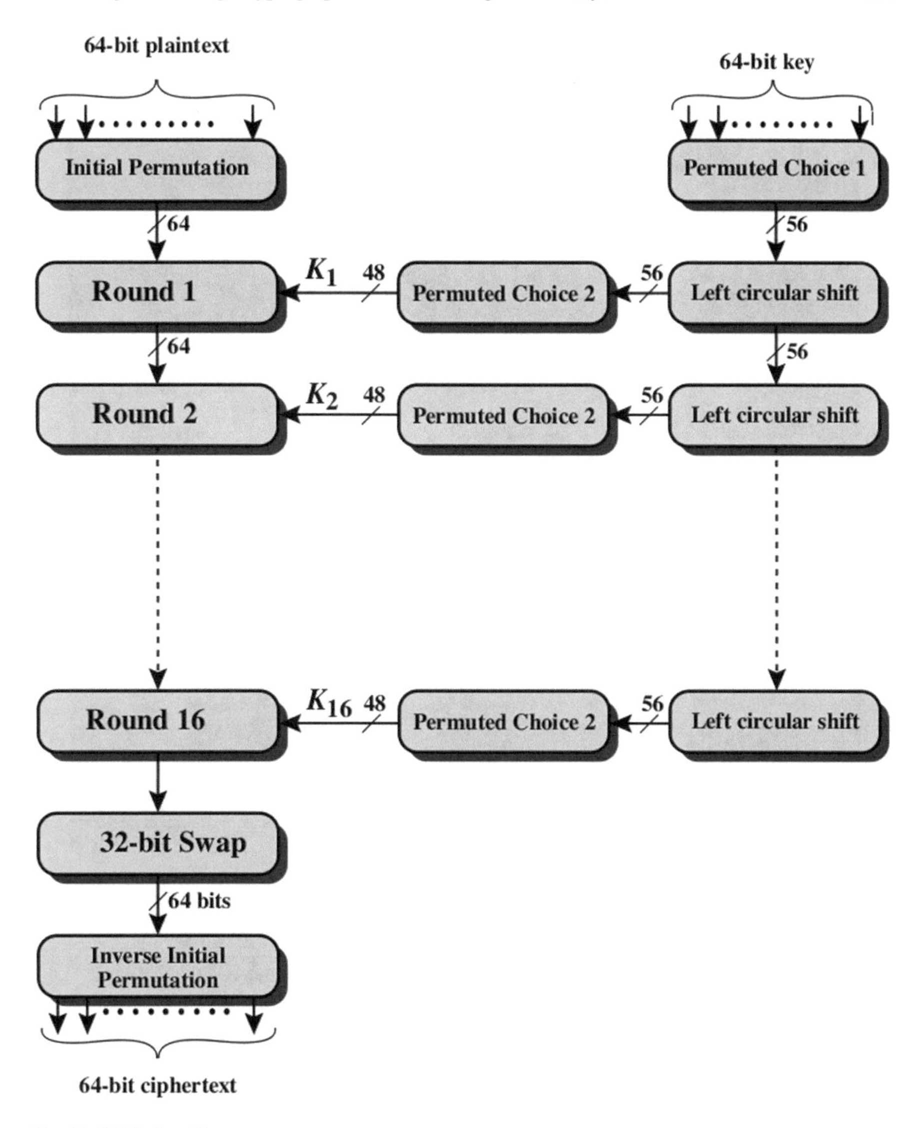

Fig. 3 DES algorithm

guess for the intruders. This provides the best ever security to big data analytics in the support to numerous data files such as media, text, and image files. The following procedure describes the process of encryption and decryption at multiple levels.

Table 3 Permutation table

(a) Initial Permutation (IP)

58	50	42	34	26	18	10	2
60	52	44	36	28	20	12	4
62	54	46	38	30	22	14	6
64	56	48	40	32	24	16	8
57	49	41	33	25	17	9	1
59	51	43	35	27	19	11	3
61	53	45	37	29	21	13	5
63	55	47	39	31	23	15	7

(b) Inverse Initial Permutation (IP^{-1})

40	8	48	16	56	24	64	32
39	7	47	15	55	23	63	31
38	6	46	14	54	22	62	30
37	5	45	13	53	21	61	29
36	4	44	12	52	20	60	28
35	3	43	11	51	19	59	27
34	2	42	10	50	18	58	26
33	1	41	9	49	17	57	25

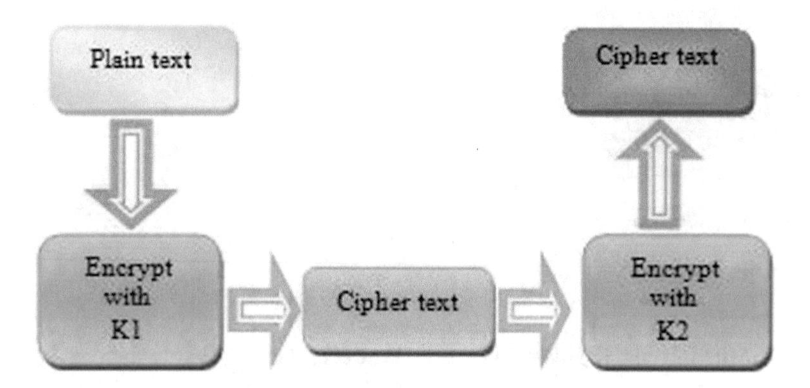

Fig. 4 Encryption with key K1 and K2

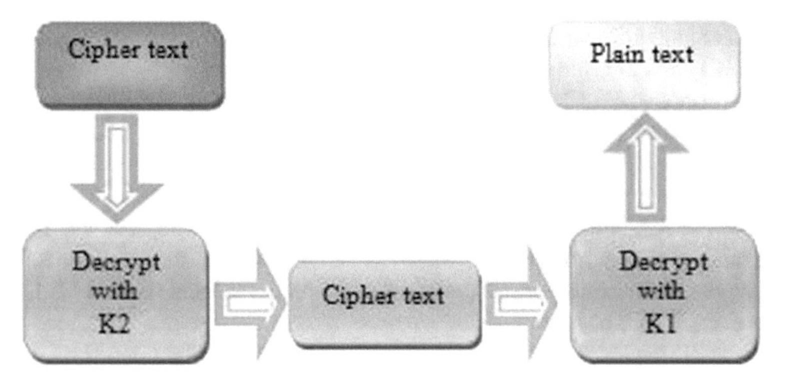

Fig. 5 Decryption with key K1 and K2

Encryption Process of MLED

1. Consider N:= number of encryption algorithms
2. Consider F: = the inputted data file.
3. Randomly sequence the N algorithms in an order from 1 to N
4. Compute a key phrase K_{ph}
5. Generate a N length random number *Rand*
6. Choose_Crypto_Algorithm ( ).

Choose_Crypto_Algorithm ( )

```
1.     Extract digit d from Rand
2.        If (d =0)
3.           then d:=d+1
4.        For i:=1 to N do
5.        {
6.           If (d= i)
7.           {
8.           then i^th cryptographic algorithm will be chosen
9.           Execute the i^th algorithm and store the output which will be supplied as input to the next algorithm
10.          Go to step 1
11.          }
12.       }
13. Store the output file (i.e. multi-cipher text) and send it via a communication link.
```

Decryption Process of MLED

1. Receive the output multi-cipher text file sent via a communication link.
2. Apply N algorithms in their reverse order of encryption process.
3. Generate the original data file.

4 Implementation of MLED

For implementation of multi-level encryption and decryption (MLED), the traditional cryptographic algorithms have been used. The algorithms such as blowfish, advanced encryption standard, and data encryption standard are combined and implemented in multiple levels to achieve more security.

In the proposed MLED model, a random function operator has been used which generates a random number. The length of the random number depends on the number of algorithms used for encryption and decryption. Since three algorithms are used in this work, there is a possibility that the randomizer can generate a 3-digit random number out of 3! Combinations, for example, 123, 231, 321, 132, 213, and 312. The 3-digit random number determines the order of execution of the algorithms. In this work, the three algorithms are assigned with the enumerations such as {1: DES, 2: AES, and 3: Blowfish}.

There are three levels of encryption and decryption in this proposed model. The output of one level will be supplied as input to the next level. Thus, the cipher-text generated by first algorithm will be supplied as input for the second algorithm [39]. Then further the cipher-text resulted from second algorithm will be supplied to the third algorithm. Finally, the multi cipher-text will be send to the receiver end through communication channel. At the receiver side, the reverse order of encryption has been followed. If order of encryption is done in order of 3-1-2, then for decryption the order will be 2-1-3. The MLED architecture is depicted in Fig. 6.

5 Experimental Result of the Cryptosystems

The time for encryption process and decryption process is taken into consideration for measuring the efficiency of different algorithms [25]. In this experiment, text files are supplied as inputs in KB and algorithms are compared with respect to the throughput.

Throughput for encryption is measured in kB/s as

$$\frac{\sum \text{input files size}}{\sum \text{ Encryption Computation Time}}$$

Similarly, throughput for decryption is defined as

$$\frac{\sum \text{input files size}}{\sum \text{ Decryption Computation Time}}$$

Encryption/decryption time and CPU processing time in the form of throughput are used as metrics to compare the performance of algorithms. The result is depicted in Table 4.

From Table 4, it is clearly visible that the throughput value of encryption and decryption process in MLED is 4.23 and 4.52, respectively. This value is minimum with comparison to the other four cryptographic algorithms considered in the experiment. Thus, the proposed MLED crypto-system performs better in terms of

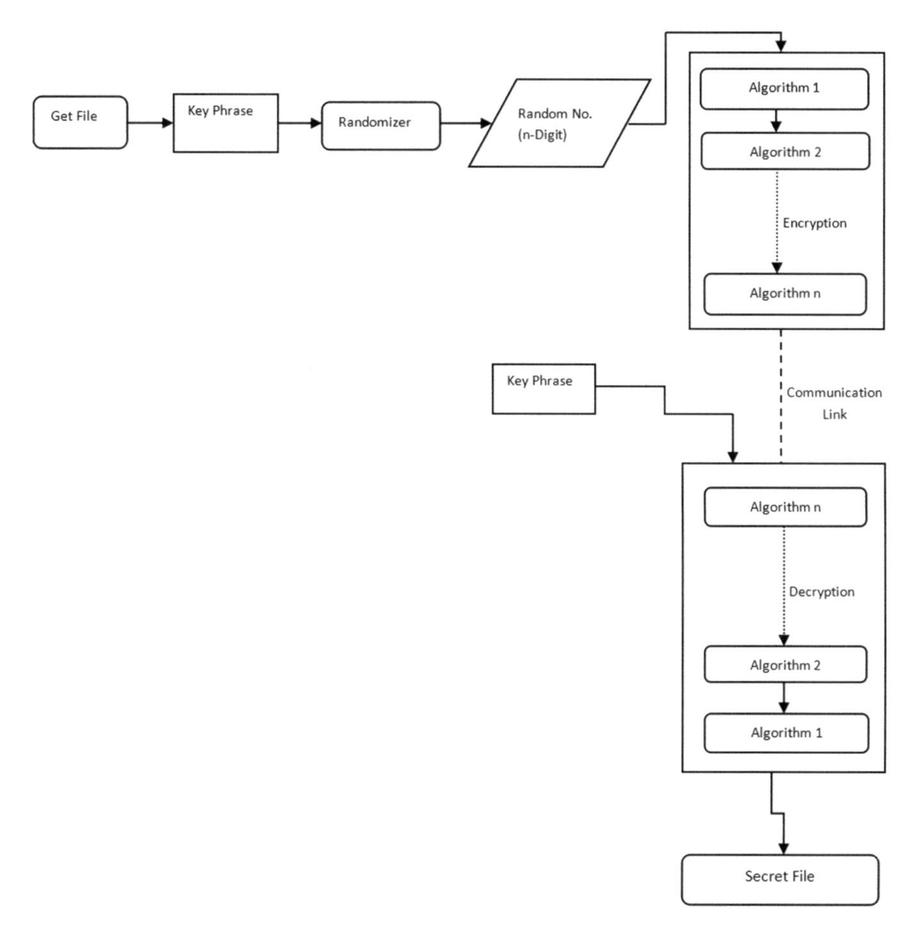

Fig. 6 MLED architecture

throughput. The encryption time and decryption time of each crypto-system are shown in Figs. 7 and 8, respectively.

The average encryption time and average decryption time of all the crypto algorithms are shown in Figs. 9 and 10.

The throughput value of encryption process and decryption process for all algorithms is compared and depicted in Fig. 11. From Table 4 and Fig. 11, it is seen that the MLED takes less CPU time for encryption process (=4.23) and decryption process (=4.52) of the whole set of inputted files.

Table 4 Processing time of all algorithms

File size(kB)	DES-Algorithm		AES-Algorithm		Blowfish-Algorithm		RSA-Algorithm		MLED-Algorithm	
	Encrypt	Decrypt	Encrypt	Decrypt	Encrypt	Decrypt	Encrypt	Decrypt	Encrypt	Decrypt
51	32	50	55	63	39	39	54	54	126	154
109	36	41	41	56	44	30	45	47	121	134
247	47	51	112	76	42	63	88	72	198	211
322	81	74	161	148	45	91	118	104	287	311
696	87	87	164	143	46	90	158	158	297	357
782	144	132	213	151	65	95	178	168	424	368
901	240	251	261	173	65	102	268	174	568	428
5501	249	242	257	171	117	99	542	383	625	437
7310	1692	1690	1365	880	105	139	961	880	3162	2960
22,300	1716	1716	1366	883	152	137	1441	961	3234	3100
Avg. Time (in Second)	432.4	433.4	399.5	274.4	72	88.5	385.3	300.1	904.2	846
Throughput (in kB/s)	8.83	8.81	9.56	13.92	53.08	43.18	9.91	12.73	4.23	4.52

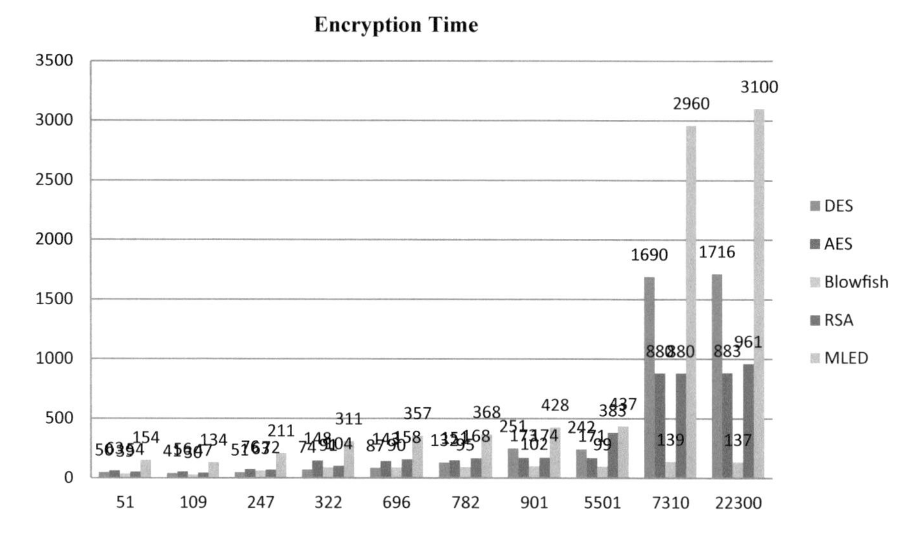

Fig. 7 Encryption time of the cryptosystems

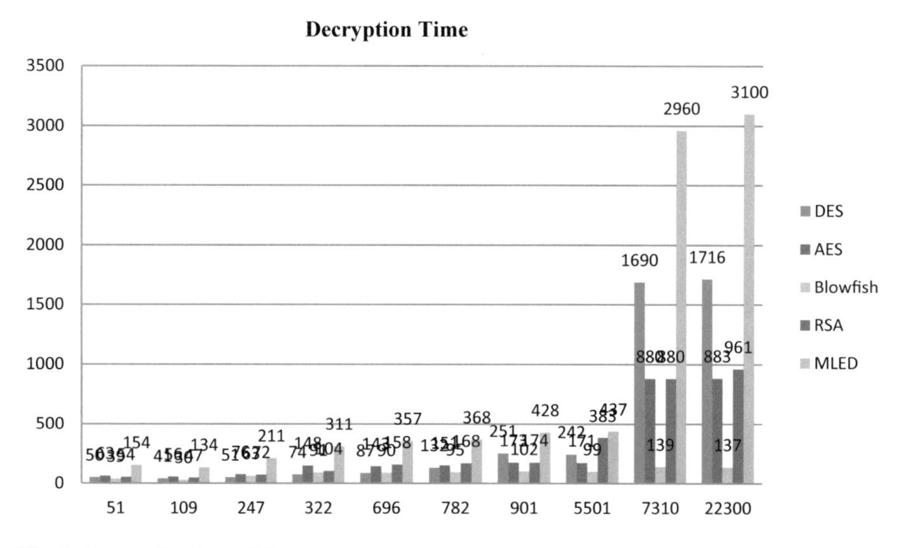

Fig. 8 Decryption time of the cryptosystems

6 Conclusion and Future Scope

Cryptographic work is implemented for privacy preservation in big data analytics. Here, the various approaches for security and privacy of big data analytics have been discussed. More security can be added to big data using crypto algorithms. The use of multilevel encryption and decryption technique is discussed thoroughly.

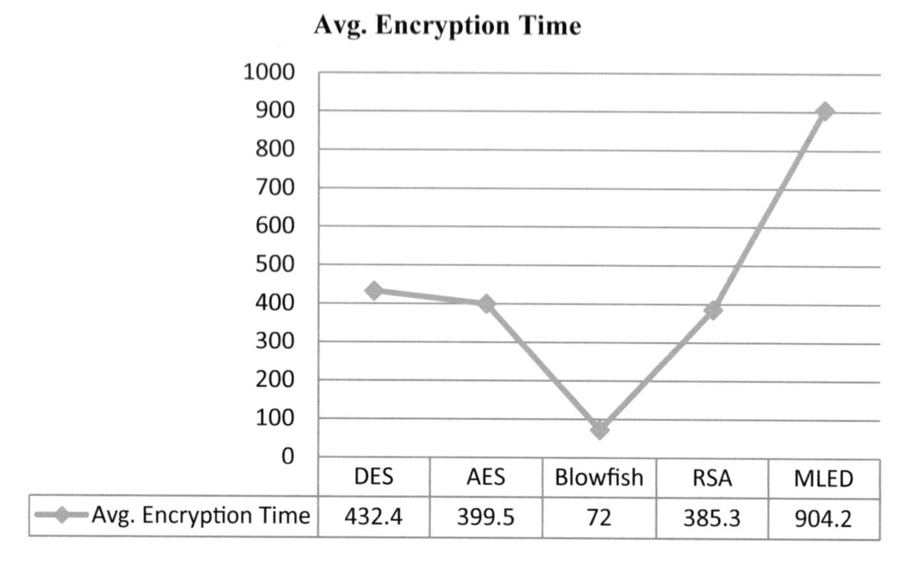

Fig. 9 Average encryption time

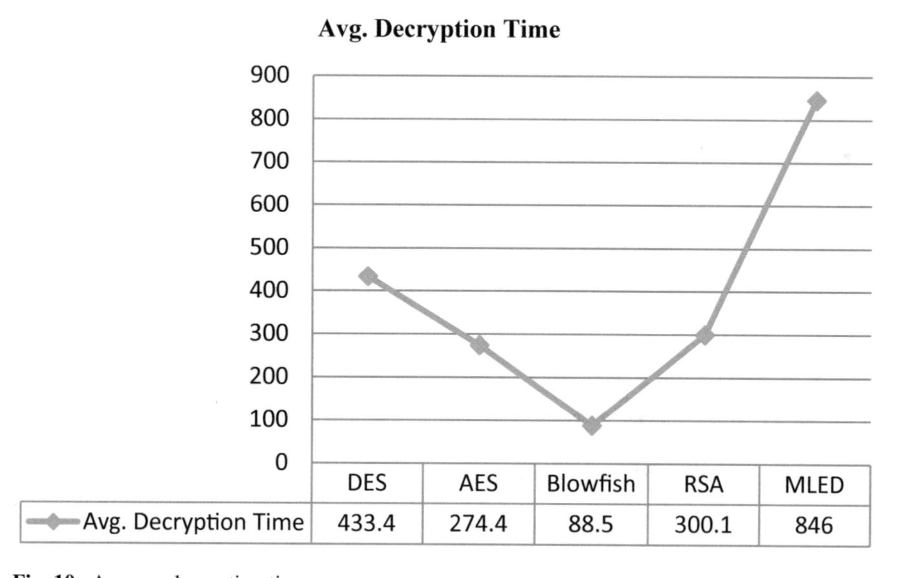

Fig. 10 Average decryption time

Also the various mechanisms which can be implemented to preserve privacy to the individual data is discussed. Each party has to encrypt data and then transmit. Then, the trusted user can decrypt and apply data analytics process. For overpowering business, big data is analyzed for extracting knowledge that lead better decisions and strategic moves. The privacy challenge in big data is a complex requirement for

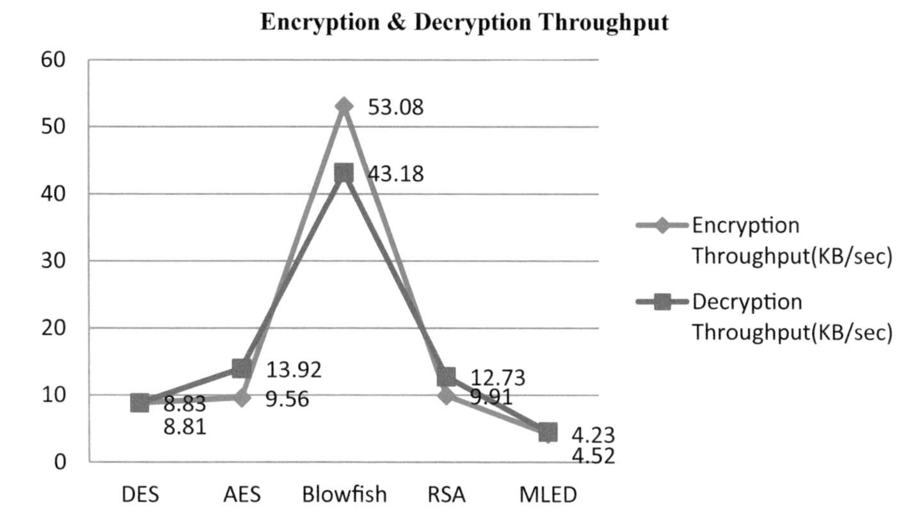

Fig. 11 Comparison of encryption and decryption throughput

privacy preservation of user's data. A novel privacy-preserving approach has been discussed into its depth for providing multiple levels of security to the sensitive personal data. Stronger encryption mechanisms can be implemented for preserving privacy in big data. It is an open research issue to address the privacy challenges. Thus, in the future, various privacy-preserving methods may be implemented. Further study on big data privacy preservation has a huge scope.

References

1. Abadi DJ, Carney D, Cetintemel U, Cherniack M, Convey C, Lee S, Stone-braker M, Tatbul N, Zdonik SB (2003) Aurora: a new model and architecture for data stream manag ement. VLDB J 12(2):120–139
2. Kolomvatsos K, Anagnostopoulos C, Hadjiefthymiades S (2015) An efficient time optimized scheme for progressive analytics in big data. Big Data Res 2(4):155–165
3. Tsai C-W, Lai C-F, Chao H-C, Vasilakos AV (2015) Big data analytics: a survey. J Big Data Springer Open J
4. Chen M, Mao S, Liu Y (2014) Big data: a survey. Mobile Netw Appl 19:171–209. https://doi.org/10.1007/s11036-013-0489-0]
5. Mehmood A, Natgunanathan I, Xiang Y, Hua G, Guo S (2016) Protection of big data privacy. In: IEEE translations and content mining are permitted for academic research
6. Kamila NK, Jena L, Bhuyan HK (2016) Pareto-based multi-objective optimization for classification in data mining. Cluster Comput 19:1723–1745. https://doi.org/10.1007/s10586-016-0643-0
7. Porambage P et al (2016) The quest for privacy in the internet of things. IEEE Cloud Comput 3(2):36–45
8. Jing Q et al (2014) Security of the internet of things: perspectives and challenges. Wireless Netw 20(8):2481–2501

9. Fei H et al (2016) Robust cyber-physical systems: concept, models, and implementation. Future Gener Comp Syst 56:449–475
10. Xiao Z, Xiao Y (2013) Security and privacy in cloud computing. IEEE Trans Commun Surv Tutorials 15(2):843–859
11. Mohammadian E, Noferesti M, Jalili R (2014) FAST: fast anonymization of big data streams. In: Proceedings of the 2014 international conference on big data science and computing, p 23
12. Evfimievski S (2002) Randomization techniques for privacy preserving association rule mining. In: SIGKDD Explorations 4(2)
13. Tripathy K, Mitra A (2012) An algorithm to achieve k-anonymity and l-diversity anonymization in social networks. In: Proceedings of fourth international conference on computational aspects of social networks (CA-SoN), Sao Carlos
14. Jain P, Gyanchandani M, Khare N (2019) Improved k-Anonymity Privacy-Preserving Algorithm Using Madhya Pradesh State Election Commission Big Data, Integrated Intelligent Computing, Communication, and Security. Studies in Computational Intelligence, vol 771. Springer, Singapore, pp 1–10
15. Kadampur MA (2008) A data perturbation method by field rotation and binning by averages strategy for privacy preservation. In: Fyfe C, Kim D, Lee SY, Yin H (eds) Intelligent data engineering and automated learning—IDEAL, vol 5326., Lecture Notes in Computer ScienceBerlin: Springer
16. LeFevre K, DeWitt DJ, Ramakrishnan R (2006) Mondrian multidimensional k-anonymity'. In: Proceedings of 22nd International Conference Data Engineering, Ser. ICDE'06. IEEE Computer Society, Washington, DC, USA, April 2006, pp 1–11
17. Zakerzadeh H, Aggarwal CC, Barker K (2015) Privacy-preserving big data publishing. In: Proceedings 27th International Conference Scientific and Statistical Database Management, Ser. SSDBM '15. ACM, New York, pp 26:1–26:11
18. Roy I, Ramadan HE, Setty STV, Kilzer A, Shmatikov V, Airavat WE (2010) Security and privacy for MapReduce. In: Castro M (eds) Proceedings of the 7th Usenix symposium on networked systems design and implementation. USENIX Association, San Jose
19. Derbeko P et al (2016) Security and privacy aspects in MapReduce on clouds: a survey. Comput Sci Rev 20:1
20. Pathak K, Chaudhari NS, Tiwari A (2012) Privacy preserving association rule mining by introducing concept of impact factor. In: 2012 7th IEEE Conference on Industrial Electronics and Applications (ICIEA), Singapore, pp 1458–1461. https://doi.org/10.1109/iciea.2012.6360953
21. Yadav GS, Ojha A (2018) Multimed Tools Appl 77:16319. https://doi.org/10.1007/s11042-017-5200-1
22. Terzi R, Terzi, Sagiroglu S (2015) A survey on security and privacy issues in Big Data. In: Proceedings of ICITST 2015, London, UK, December 2015
23. Kacha L, Zitouni A (2017) An overview on data security in cloud computing. In: CoMeSySo: cybernetics approaches in intelligent systems. Springer, pp 250–261
24. Rath M, Mishra S (2020) Security approaches in machine learning for satellite communication. In: Machine learning and data mining in aerospace technology. Springer, Cham, pp 189–204
25. Mishra S, Mallick PK, Jena L, Chae GS (2020) Optimization of Skewed data using sampling-based preprocessing approach. Front Public Health 8:274. https://doi.org/10.3389/fpubh.2020.00274
26. Ilavarasi K, Sathiyabhama B (2017) An evolutionary feature set decomposition based anonymization for classification workloads: privacy preserving data mining. J Cluster Comput (Springer, New York)
27. Jena L, Kamila NK, Mishra S (2014) Privacy preserving distributed data mining with evolutionary computing. In: Proceedings of the International Conference on Frontiers of Intelligent Computing: Theory and Applications. Advances in Intelligent Systems and Computing, vol 247. Springer, Cham. https://doi.org/10.1007/978-3-319-02931-3_29
28. Mishra S, Tripathy N, Mishra BK, Mahanty C (2019) Analysis of security issues in cloud environment. In: Security Designs for the Cloud, Iot, and Social Networking, pp 19–41

29. Sweeney L (2002) k-anonymity: a model for protecting privacy. Int J Uncertain Fuzziness Knowl Based Syst 10(5):557–570
30. Mishra S, Mahanty C, Dash S, Mishra BK (2019) Implementation of BFS-NB hybrid model in intrusion detection system. In: Recent developments in machine learning and data analytics. Springer, Singapore, pp 167–175
31. Mishra S, Tripathy HK, Mallick PK, Bhoi AK, Barsocchi P (2020) EAGA-MLP—an enhanced and adaptive hybrid classification model for diabetes diagnosis. Sensors 20(14):4036
32. Bredereck R, Nichterlein A, Niedermeier R, Philip G (2011) The effect of homogeneity on the complexity of k-anonymity. In: FCT, pp 53–64
33. Jain P, Gyanchandani M, Khare N (2019) Enhanced secured map reduce layer for Big Data privacy and security. J Big Data 6:30. https://doi.org/10.1186/s40537-019-0193-4
34. Mishra S, Mishra BK, Tripathy HK, Dutta A (2020) Analysis of the role and scope of big data analytics with IoT in health care domain. In: Handbook of data science approaches for biomedical engineering. Academic Press, pp 1–23
35. Mishra S, Tripathy HK, Mishra BK, Sahoo S (2018) Usage and analysis of Big Data in E-health domain. In: Big Data management and the Internet of Things for improved health systems. IGI Global, pp 230–242
36. Mishra S, Tripathy HK, Mishra BK (2018) Implementation of biologically motivated optimisation approach for tumour categorisation. Int J Comput Aided Eng Technol 10(3):244–256
37. Shen Z, Li Li, Yan F, Xiaoping Wu (2010) Cloud computing system based on trusted computing platform. Int Conf Intelligent Comput Technol Autom 1:942–945
38. Mishra S, Sahoo S, Mishra BK (2019) Addressing security issues and standards in Internet of things. In: Emerging trends and applications in cognitive computing. IGI Global, pp 224–257
39. Kaur M, Mahajan M (2012) Implementing various encryption algorithms to enhance the data security of cloud in cloud computing. VSRD Int J Comput Sci Information Technol 2:831–835

Chapter 8
Application of Big Data Analytics in Healthcare Industry Along with Its Security Issues

Arijit Dutta, Akash Bhattacharyya, and Arghyadeep Sen

1 Introduction

The challenges in the information technology department have grown their roots in the healthcare system. This has resulted in poor healthcare information and data management. There is a high volume of data transfer, and the variety of the data directly affects the complexity of the processing. This results in an increase in the costing and time consumption in the processing of a healthcare provider organization. This has made the organizations to seek solutions and processes, which would help them in the processing of the data in a more effective and data-driven model for their business [1].

Data scientists have been able to work on the processing of big data. Big data analytics has been found to be able to process of the 3Vs of big data—Volume, Variety and Velocity. Big data analytics have been found to be one of the most important aspects of the working of healthcare data [2]. Research has found that the use of big data analytics has been lagging in the healthcare industry when compared to other sectors like banking and retailing industry. The healthcare industry has been trying to implement the analytics in their processing with the proper investment in the technologies [3, 4]. Based on a survey, it has been found that only 40% of the total healthcare industry has proper use of analytics in their possession. This low data is due to the inability to understand the importance of the use of analytics. Out of the 40%, only a small percentage of 16% are able to understand the proper working of analytics in their system. There is a high urgency in the healthcare industry to make them understand the economic and the strategic impact on the processing of the data, which big data analytics will have on their organization [5].

It has been found that the main goals for the completion of this study are to identify the prospective capabilities big data analytics would have on the healthcare processing and industry and to explore the different beneficial aspects it would have

A. Dutta (✉) · A. Bhattacharyya · A. Sen
School of Computer Engineering, KIIT Deemed to be University, Bhubaneswar, India

P. K. Das et al. (eds.), *Privacy and Security Issues in Big Data*, Services and Business Process Reengineering, https://doi.org/10.1007/978-981-16-1007-3_8

on the current industrial scenario. In this paper, there is a historical background of the development of big data analytics in healthcare and moving towards the research study of different cases. The use of resource-based study has been conducted for this research work. The study of the cases has helped in the studying of the strategies, which can be used by the healthcare industry, and some of the potential applications where big data analytics have been able to provide proper application window to researchers. The study would help in the understanding of the big data analytics value in the processing of the business and provide with a proper guidance with the inclusion of evidence for the healthcare industry. Following this, a presentation of the limitations of this study and the future research work has been compiled.

2 Background

2.1 History of Big Data Analytics

The term big data was coined by Michael Cox and David Ellsworth in the year 1997 in a paper [6], which discussed about the data visualization and the drawbacks, which it held for the computer systems. The 1990s had enabled the rapid growth of large amount of data alongside the IT industry. However, there was less amount of proper usable information from the data that was generated. The idea of business intelligence has started to take its peak in the industry for the collection and analysis of the data being generated. This processing would help the industries to take proper decisions in their respective departments. In the 2000s, big data got a revolutionary outbreak. The definition of the big data was made with the help of 3Vs: Volume, Variety and Velocity. This helped in the development of proper software, which was able to process the high amount of data and provide the required amount of information from them. Big data analytics is said to be a double-edged sword, collection of high financial cost due to poor governance, whereas the proper use of the tools would help in the processing of huge amount of data and provide proper medical support at a lower cost. Indefinite of the factors, the healthcare industry thrives to survive in the corporate war and is unaware of the huge prospect of big data analytics has on the data that is being generated. Proper understanding of the benefits of big data analytics needs to be provided to the healthcare industry, which would facilitate the integration of the processing capability into their system. This processing would help the healthcare participators and practitioners to take control of the big data analytics in their industry. The data collected from different locations was in different formats and was being generated at a high rate, which was more than the speed at which data being processed. The high amount of data forced the healthcare industry to follow proper formats for the storing of data digitally. By 2009, big data had been found to be of great benefit for the researchers [7] as well as organizations who were using the process for their decision-making system. However, the use of proper big data analytics in the healthcare industry is for the benefit of having predictive modelling in the tools and system. These predictive models have not yet been developed by the researchers for the use in healthcare industry.

2.2 *Architecture of Big Data Analytics*

For the research to be completed in accordance with the goals, the proper understanding of the big data analytics architecture needs to be known. The big data analytics architecture is in accordance to the data life cycle, which helps in the processing of data collected from various sources. The architecture has five major layers as shown in Fig. 1.

2.2.1 Data Layer

The data layer consists of all the data sources, which are necessary for the provision of insights to the data, which are collected and would thus help in the processing and solving of various business problems. Data is collected from various internal and external sources and is then stored in respective database based on their format: structured data, semi-structured data and unstructured data.

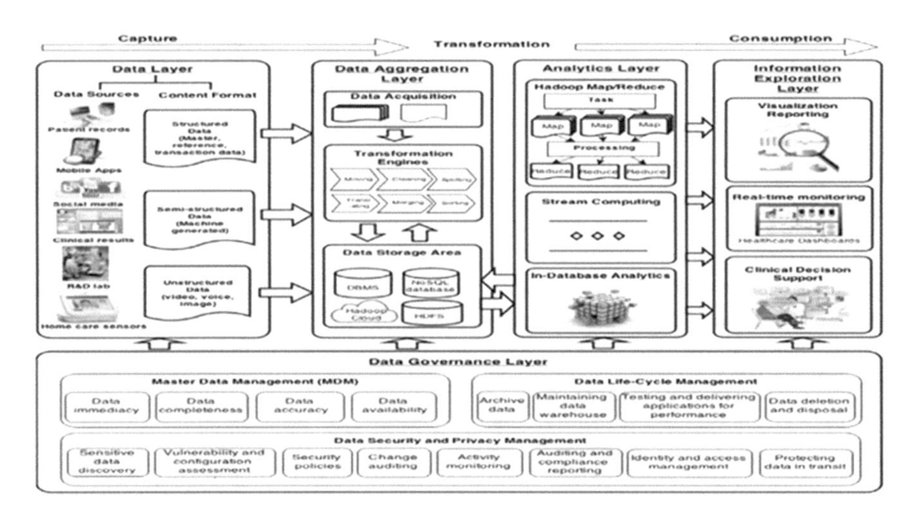

Fig. 1 Big data analytics architecture in health care (*Source* [8])

2.2.2 Data Aggregation Layer

All the data collected from various sources are now processed to be used in the analytics processing. The data collected is cleaned and found to be appropriate for the analytics or not. The data is then transformed into the format, which would be helpful for the softwares to read. The transformation processing must be able to extract process and validate the data for the storage processing. The data is eventually converted into a required format and stored in Hadoop distributed file system (HDFS). The data processing techniques can be processed in real time or can be batch controlled.

2.2.3 Analytics Layer

This layer is considered responsible for the working on the data that has been collected and stored. For the proper working on the data, there is a requirement of tools, which would be able to work on the data. These tools are found in the analytics layer. One of the most commonly used analytics tools is the Hadoop MapReduce technique. This tool is able to process large amount of data with ease. The inclusion of real-time data analytics helps in the tracking of the data being generated. This real-time tracking can be used for finding fraud in the data stream. Any abnormality in the data stream would raise an alarm for the corresponding data. This would help the organization to catch major health frauds from their system. The analytics layer is able to provide the use with a high exception supported evidence-based medical reports.

2.2.4 Information Exploration Layer

The layer is responsible for the use of the information generated by the previous layer and transforms it into visualizations, which would be easier to understand than the numerical tabular data. The development of proper report is an integral part of any data analytics process. Considering the healthcare industry, it has been found that the most important tracking method of the data can be in the real-time tracking methodology.

2.2.5 Data Governance Layer

This layer helps the organization to harness the data that is being generated by them. The inclusion of master data management module helps in the management of the data. Management of data over a long period requires proper access to the data security as well as the privacy management of the data. The data that is being generated and stored also needs to follow the data life cycle, which helps in the maintenance of the flow of data in an organization. Due to the increase in the complexity of privacy and security, the organizations often face legal issues related to data theft and data

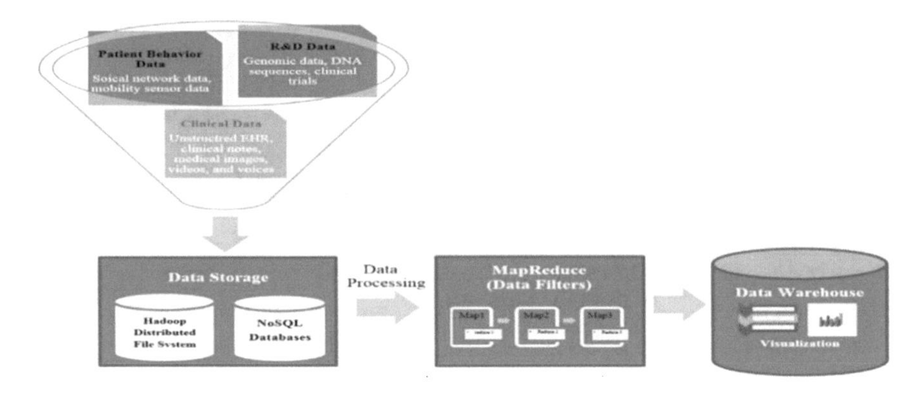

Fig. 2 Analytical processing of unstructured data in healthcare industry (*Source* [11])

loss. Using suitable policies and compliance requirements, these drawbacks can be reduced.

2.3 Capability of Big Data Analytics

Out of several definitions created for the capability of big data analytics, the general definition says about the ability of a system to manage a large amount of information and data in order to allow the users to implement data analysis [9]. Researchers have indicated that big data has the capability for the maximization of business value to be able to provide faster insight into the data and the ability to change the raw data into usable visualized information. From the analytics point of view, it has been stated that the big data analytics is divided into three categories: aspirational, experienced for a business operation optimization and transformed for making a targeted investment possible [10]. In the context of healthcare, big data analytics capability can be defined as the process, which can provide the user with the power to collect, study and analyse a large amount of health-based data and to provide proper insight into the data on time as explained in Fig. 2.

2.4 Conceptualization of the Potential Benefits in the Use of Big Data Analytics

In order to capture the potential benefits in the use of big data analytics, various benefits were collected from the cases collected. The benefit designed includes IT infrastructure, operational benefits, managerial benefits, organizational benefits and strategic benefits [12]. This framework has been designed as per the Shang & Seddon's framework for the classification of the potential benefits collected.

Table 1 Designed framework

Dimension of benefit	Description	Sub-dimension
IT infrastructure benefits	Development of IT resources which would help in the foundation of future application of business	Flexible business building along with future changes
		Cost reduction
		Increase in the infrastructure capabilities
Operational benefits	Operational activities have a direct effect on the benefits of the organization	Cost management and reduction
		Reduction in cycle time
		Improvement of productivity
		Improvement of quality
		Customer service management
Managerial benefits	Monitoring and controlling of operations provide beneficial outlook in an operation	Resource management
		Improvement in decision-making system
		Boosted performance
Strategic benefits	Benefits in the planning of long-range planning and decisions	Growth of business
		Development of business alliance
		Innovations in business
		Low-cost leadership
		External allied linking
Organizational benefits	Use of an enterprise resource system helps in the benefit of the organization	Work pattern modification
		Organizational learning
		Employee empowerment
		Common vision enforcement

This framework has been considered for the completion of this study as the study conducted is mostly explanatory which helps in the understanding of the concept. The design of the framework is such that the managers use this to assess their organisation enterprise system. The framework designed is shown in Table 1.

3 Research Methodology

A qualitative study had been conducted for the completion of the study. Multiple cases had been considered for the gaining of knowledge and understanding of the big data analytics in health care and the capabilities and corresponding benefits from the use of analytics. The collection of data and analysis has been discussed in the following sections.

3.1 Case Collection

The case studies were collected based on two selection criteria: having real-time implementation of big data analytics tools in health care and the techniques use and the potential benefit of the use of big data analytics. Big data cases of major IT companies had been collected from various sectors and assessed. Cases having no industrial link and vague descriptions had been removed from the cases considered. The final set of cases is comprised of 100 cases.

3.2 Research Approach and Procedure

Three-phase data processing had been considered for the completion of this study. Research had been compiled in the form of content analysis. Data was collected based on case reading. Content was extracted from various themes. The three-phase (preparing, organizing and reporting) study helped in the covering of all the cases collected and ensured better understanding of the benefits and capabilities of big data analytics.

- Preparation phase was collection of any related information from the cases, which were in the form of sentences, or a portion of a page. Manual selection of textual content described the benefits of big data analytics in healthcare industry. Data was collected and stored in excel sheet for further investigation.
- Organizing the data that had been collected was done with the help of creation of categories and classification techniques. The data was grouped based on conceptual theme similarity. This helped in the reduction of the data into compact categories of data having similar meaning. This process was conducted twice having a broad knowledge of big data applications and not considering the previous result collected. This resulted into two broad categories being created based on the data collected.
- Reporting of the data had been conducted based on the classification of the data in the previous phase. The patterns were subdivided into five separate groups based on the capability of having big data analytics in health care. The potential benefits of the same were considered as one broad category of classification.

Capability of big data analytics can be classified into sub-categories having analytical processing of care delivery, unstructured format of data analytics, and provision of a decision support system for the user, prediction of future events and traceability of certain aspect of health care [13–15]. Potential benefits have been considered based on the Shang & Seddon's framework for organizational benefits.

4 The Strategies to Get Success in Big Data Analytics

In order to implement a big data-driven environment in an organization, the stakeholders need to identify the business value of using big data analytics in their environment. They also need to understand the benefits, which they would be able to achieve with the help of big data analytics. Many of the organizations are unaware of these facts. Most importantly, the healthcare industry has been kept in the dark for a long period with the non-integration of big data analytics in their sector. This research has thus been conducted to help these organizations to get an overview of the strategies and the implementation process in their sector. Accordingly, they would also be able to understand the benefits of using big data analytics in their sector. The following discussion would help the reader to understand the importance and the areas in which big data analytics can be useful for the healthcare industry.

4.1 Importance of Big Data Analytics in the Healthcare Industry

For the process of understanding the implementation process, it is important to first understand the importance of having big data analytics in their organization. The following list discusses the importance of having big data analytics in the healthcare industry:

- Patient-related services: In order to provide faster relief for the patients, big data analytics can be used to provide solutions for the medicine-based diseases at the initial levels of clinical autopsy. This can help in the minimization of drug overdosage and provide efficient medicine for diseases. This in turn helps in the reduction of readmission and thus in the cost reduction of the services [16].
- Detection of spreading-based diseases: The prediction of virus-affected diseases can be conducted with the help of live analysis of the disease. The analysis of the geo-location-based study of the patient can help in the provision of remedies to the affected patients. This can also help the professionals to provide faster remedies to the patients when they arrive.
- Hospital quality management and monitoring: The analysis of the hospital set-up according to the advised norms is an integral part of any hospital set-up. The analysis would be able to help the hospital authorities to control the patient relationships better [16].
- Improved treatment methods: Disease-based customized treatment can be implemented with the help of data collection from the patients of the hospital. This can help the doctors to administer the required dosage of medicine at a faster pace. Monitoring of the dosage provided and the changes in the dosage can be analysed to understand the correct dosage of the medicine, which is required for the eradication of the disease.

4.2 *Big Data Ecosystem Designed for Healthcare*

In order to implement the big data architecture in the healthcare system, there needs to be the proper understanding of the requirements of the organization. This can be done with the help of the following architectural design of the ecosystem, which can be implemented in a healthcare industry as explained in Fig. 3.

The authorities for the patients being admitted in the hospital generate electronic health records. These data are then transferred to the HDFS with the help of flume and sqoop. HIVE and Map-Reduce are used for the analysis of the data being collected. Similar patterns are found out from the data being collected which helps in the modelling of the predictive model. The predictive model would be able to study data inputs and provide risk-reducing methods for the disease being diagnosed. Storing of the multi-structured data is done using HBase [16]. Live analysis and informing of any emergency abnormalities can be conducted with the help of STORM. Collective data and analysis report are generated with the help of Hunk and Intellicius.

5 Potential Application Areas

The above sections are discussed about the application area for smaller integration of big data analytics in a healthcare organization. The process can be used to provide remedies and analyse data for smaller diseases. This section discusses the process of data for the use in large-scale data and diseases, which require much more analysis than that of simple data. These applications are currently being studied to make more robust in nature.

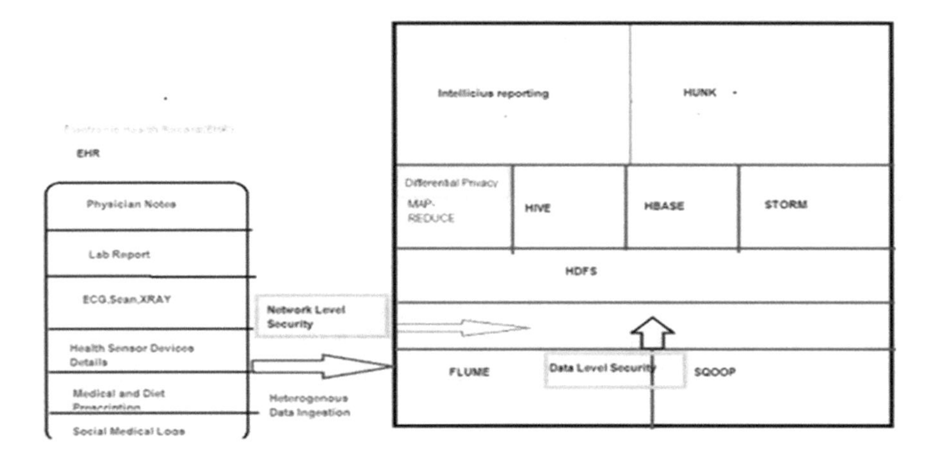

Fig. 3 Big data healthcare architecture (*Source* [16])

5.1 Big Data Used for Medical Image Processing

Imaging of various organs helps in the detection of diseases states as well as adds to the knowledge of the understanding of the organ. Medical imaging has been used for identification of lung tumours, delineation of organs, deformity in the spine and so on. This study understands the image processing procedures like enhancement of the image, segmentation and noise reduction with the help of machine learning methods [17]. With the rise of data quantity, there is an increase in the complexity of the analysis. This requires larger and more equipped system. The rise of patients in the world has prompted the use of computer-aided diagnostic solutions as well as decision support system. Many areas in medical diagnosis can be greatly improved with the help of computer intelligence. Use of analytics has the power to improve the diagnostic accuracy. Integration of medical image processing with the help of electronic health record can be used to improve the accuracy as well as the time taken for the completion of the diagnosis.

5.1.1 Image Processing Data

For different medical methodologies, there are various types of imaging processing with a wide array of processing capability. For example, visualization of blood vessels in a human being, computed tomography (CT) or magnetic resonance imaging (MRI) could be used [18]. Looking at the data that is being collected, the data has been found to be in different dimensions (two, three and four). Functional MRI and positron emission tomography have multidimensional medical data values. Modern medical imaging equipments are capable of producing high-quality images [19, 20]. This high volume and variety of data in the medical industry are making the process of analysing them a great challenge. Advancements in the medical data analytics would be able to provide with personalized care along with predictive modelling of disease and decision-making system for the diagnostics being carried out. There are two methods of imaging process, which are used for the process of medical imaging:

- Molecular Imaging: A nonintrusive process of cellular event, which has the main application area such as cancer.
- Microwave Imaging: A process of creating electromagnetic wave for the creation of a map based on the dielectric property of tissues.

The integration of different images from the processes conducted would be useful in order to improve the diagnostic accuracy as well as designing the predictive model of disease detection. However, there is a challenge of finding a proper storing space for the data that is being collected, analysis of the data being collected as well as designing of a mapping and dependency of various types of data that are being stored. These challenges are yet to be resolved.

5.1.2 Methodology

There has been an exponential growth of medical images. Apart from the volume of the images being created, the images are of different resolutions and dimensions. These problems give rise to data integration procedure as well as use of proper data mining algorithm over multiple data sets. There are a considerably less number of multimodal medical image analysis than the single-modal image analysis [21, 22]. When the data is being used at the institutional level, there needs to be the proper testing of the system to be able to work on the data. The designing of a good annotated data structuring method is a great challenge. This challenge increases when the data from various other intuitions are incorporated simultaneously. Different institutions might design their system differently, which makes the process of integrating everything together into a single system a greater challenge for the analyst. For the process of incorporating the multimodal image processing, there needs to be the use of a real-time feasible and scalable system.

Analytical Methods: The objective of use of medical image processing and analytics is for the improvement of the interpretability of the contents of the images. Various methods and frameworks have been designed for the processing of medical images. However, there is no information, which suggests that these methods are applicable for big data processing applications [23]. One such framework, which has been developed for the analysis of large data set, is Hadoop, which incorporates MapReduce. MapReduce framework is currently being used for three main cases:

1. Use of support vector machines for the lung texture classification.
2. Medical image indexing based on the content.
3. Solid texture using classification process to analyse wavelets.

Designing a system, which would be able to analyse the data at a faster rate, is important to be utilized and deployed in critical conditions. The system should be able to analyse the data collected and provide correct suggestion in critical conditions. This means that the real-time execution complexity should be as low as possible. Another factor, which needs to be kept in mind, is the accuracy of the results [24, 25]. Incorporation of dependency factor among a large number of attributes in the dataset would be helpful for the improvement of accuracy. A decision support system had been designed which was able to provide accurate treatment decision for the patients suffering with traumatic brain injury (TBI). Demographic information, features and medical records had been extracted from the CT scans conducted to analyse and predict the intracranial pressure on the brain. The accuracy had been found to be around 70% [26, 27].

Collecting, Sharing, and Compressing Methods: Along with the process of development of analytical processing methods work is being done in order to collect, compress, share and anonymize medical data [28]. Implementation of a singular centre for the collection of biomedical data from various sources and sharing the data with protected privacy is being done with the help of iDASH [29]. iDASH is known as integration of data for analysis, anonymization and sharing. Based on Hadoop ecosystem, a system has been developed in order to exchange, store and share electronic records among different healthcare institutional systems. This system is capable of storing medical images from a collective database. The medical data has been collected from a group of patients with the help of various sensors and imaging equipments. The system makes use of cloud storage platform for the data collected. A system has been designed and a prototype has been implemented which works for the storing, execution of query and collection of data requests on a database of Digital Imaging and Communications in Medicine (DICOM). The system makes use of Microsoft Azure as the cloud platform.

There are certain limitations to the use of the application specific compression methodology in both general processors and parallel graphics processing units. This is due to the inclusion of algorithms with high variable controls as well as complex manipulation in bits, which does not work well with pipeline and parallel architecture. To overcome the challenges, an implementation was proposed which uses LZ-factorization and directly decreasing the computational burden of the algorithm [30]. These are some of the techniques, which are being designed for the use in limited application processes. It has been found that the use of higher processing and analysing technique and on a large scale of data with a high accuracy is still a critical task.

5.2 *Medical Signal Analytics*

Physiological signal and telemetry monitoring devices are all around us. However, the amount of data being generated has not been stored for a long time. This makes a challenge in the extensive investigation process. The implementation of the analysis of the data being collected has started in recent times, which is helping in the improvement of patient care and proper management in healthcare industries. Analysis of streaming data has been conducted with the help of continuous waveforms along with related medical records. These forms of analysis help in the process of development of a decision-making system, which can be used to provide prompt healthcare help to the patients [31] as shown in Fig. 4. To complete such challenge, there needs to be the use of a platform, which would be able to collect and store streaming data of a high bandwidth and multiple waveforms. Implementation of the waveforms being collected needs to be mapped to the EHR data of the corresponding patients. This would help the analytical engine to have contextual and situational awareness. The analytics, which can be conducted on the data, can be either prescriptive, diagnostic or predictive. These analytics can be used to design corresponding mechanisms and systems, which can send messages and notification to the concerned person.

A challenging task in process would be to synchronize the data that has been collected from the patients and the patients' information in order to create the next generation of treatment. The following sections discuss the challenges and the developed system for the monitoring of high fidelity data as well as non-continuous data [32].

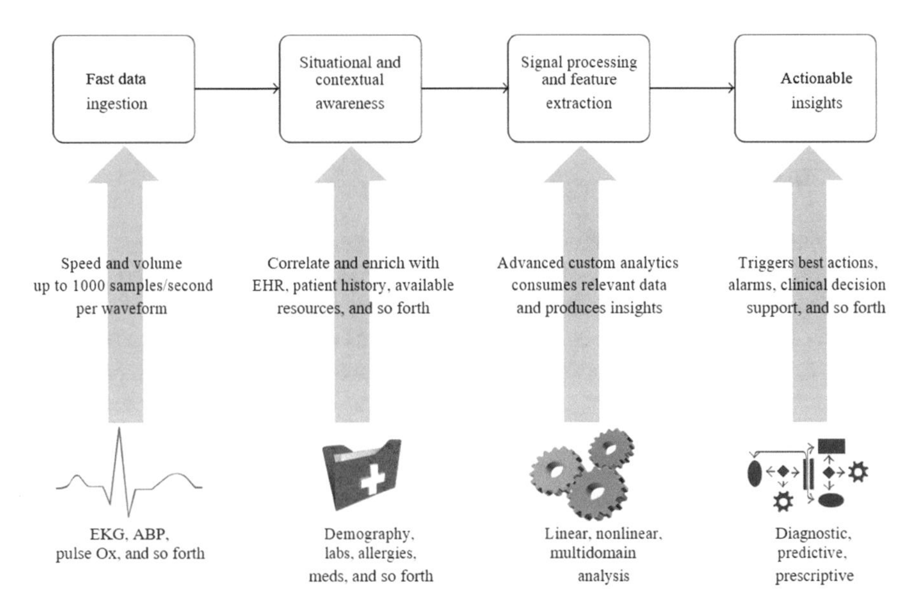

Fig. 4 Analytical workflow for the streaming of healthcare data (*Source* [32])

5.2.1 Data Acquisition

Historical data from continuous data streaming signal devices was seldom stored. Even if the data was stored, the data was small and could only be used with the help of proprietary software from the equipment manufacturer. In recent times, many of the developers of such software and hardware are providing interface, which can be used for streaming live data streams from the devices. The challenge with such process would be the provision of high bandwidth, scalability of the system and the initial cost for the set-up of the system [33, 34]. These challenges are stalling the process to be adopted into the healthcare system.

5.2.2 Data Storage and Retrieval

For the simplicity in usage procedure, there has been the rise in the acceptance of systems like MapReduce, Hadoop and MongoDB in the healthcare community. MongoDB is a cross-platform data with document-oriented database [35]. This reduces the use of traditional relation-based tables. Every health system has their dedicated database system, which creates a challenge for the healthcare system of multi-institutional data or in case of research studies. Moreover, the integration of streaming waveform data from various devices and the EHR data is not collectively operable. This is the challenge, which MongoDB, can help with the provision of various performance, scalability and availability of data. These technologies also help researchers to work on the same data in order to find scientific discovery for the process of finding suitable solutions for the healthcare industry [36].

5.2.3 Data Aggregation

Data collected from the medical industry is complex in nature. They can be both interdependent and interconnected with one another. Medical data also has the inclusion of various security and privacy requirements from the governing bodies of data. The data thus needs highly secured storage as well as privacy-controlled access with proper following of data sharing ethics. Continuous data analysis utilizes proper time frame domain. However, in case of static data, there would not be any inclusion of time variable data [37, 38]. This problem gives rise to the challenge when the integration of time-sensitive waveform data is connected with static EHR data. Considerable amount of effort is included in the process of integrating time-sensitive data. The time-sensitive waveform data is collaborated with collaborated with static EHR data into a single cohesive database, which is made public for research.

5.2.4 Signal Analytics Using Big Data

Intensive care units (ICUs) of hospitals produce vast amount of data in a short period. Data is collected from all the patients in the hospitals ICU. A scalable infrastructure has been proposed to be implemented in the ICU of hospitals, which would incorporate live stream data collected from the sensors and analysed in real time. The current research in the field incorporates the data from the heart and respiratory sensors to extract biomarkers from signal processing. This system should be able to maintain a stable heart rate and respiratory variability with the help of spontaneous breathing trends [39]. Research has been done on the data that is collected from both preoperative and post-operative, which is used to monitor the status of the patients. The potential use in the development of machine learning models infused with data analysis using biomarkers has been implemented.

Neurological data analysis on patients has been able to help in the understanding of various complex diseases as well as help in the production of therapeutic and diagnostic devices which would be able to help patients. A research has been conducted on neuro-critical care for patients with the use of various monitoring systems [40]. A multimodal monitoring system for traumatic brain-injured patients has also been implemented which provides patient-centred care. With the rise in complex physiological monitoring systems, the price has been reduced and patients are embracing them outside the hospital. However, with large number of devices, the data to be collected and stored can be a great challenge. Similarly, the use of portable echocardiogram machines, body weight monitoring devices and blood pressure devices are being developed to be set up in a network environment [41].

5.3 Genomics: Emerging Big Data Applications

The advancement of high-powered sequencing methods has helped researchers to study genetic markers from a large population. This research study has been able to increase the magnitude of the study of human genome by a factor of five folds [42]. This study can also help in the application of associated generic causes in disease states. Genome-based analysis has been able to prove helpful for the similar application in trait study of the population as well as in finding treatment for chronic and complex diseases.

Analysis of the genome sequencing problem is a big data problem as a human genome is made up of around 35,000 genes. Implementation of integration of timescale clinical genomic data is to be implemented into the physiological level data collected from humans. This would help in providing personalized care to each patient as per the genes the patients consist of [43]. In order to deliver reliable and correct recommendation to the patients, the system needs to be reliable. This research field is still under speculation with primary objective in the research of cancer genes. This has been such a scenario due to the high cost incurrent, time-consuming process as well as a huge labour-intensive analytical problem-solving research study.

Application of big data analytics in the genomics field has incorporated a large number of topics over the recent few years. The main focus has been made on pathway analysis. The functional effects of various genes are expressed differentially in experiments, or gene traits of a singular interest is analysed, and a network is reconstructed. The network is then tested and analysed with the help of different variables. This process helps in the understanding of various cellular processings in the human body [44].

5.3.1 Pathway Analysis

The resources for the inferring of the functional effects of genomics are predominant on the use of statistical association among the gene changes which have been observed and the changes which have been predicted. Experimentation and analytical processing of the data lead to the generation of error as well as batch effects. Functional effect interpretation needs the incorporation of a constant increase in genomic data along with a corresponding gene annotation. There has been found to be the availability of various high standard functional pathway analysis data on genomic levelled data [45]. There are three generations of pathway analysis methodology as described below and shown in Table 2.

- First Generation: It consists of the overrepresentation analytical approach which helps in the determination of a small fraction of genes from a singular pathway collected from a gene which had been incorrectly expressed. Some of the better-known first-generation tools available are Onto-Express, ClueGo and GoMiner [46].

Table 2 Toolkit examples and application domain

Toolkit name	Category	Selected applications
Onto-express	Pathway analysis	Breast cancer
ClueGo	Pathway analysis	Colorectal tumours
GoMiner	Pathway analysis	Pancreatic cancer
GSEA	Pathway analysis	Diabetes
Pathway-express	Pathway analysis	Leukaemia
ODE models	Reconstruction of gene regulatory networks	Cardiac development
Boolean methods	Reconstruction of gene regulatory networks	Cardiac differentiation

- Second Generation: This includes the use of functional class scoring methodology, incorporating changes in the expression level in individual genes along with functionally alike genes. GSEA is one of the popular tools used for second-generation pathway analytical processing.
- Third Generation: It makes use of pathway analytical topology generating tools. They are publicly available knowledge infused database with a detailed information about the gene interactions. The interactions recorded are the location and the process of the interaction of two genes.

5.3.2 Reconstruction of Regulatory Networks

It has been found that the pathway analysis approach is not able to help in the processing of high-throughput big data in the biological research area without the use of integrated operation of dynamic systems. There are various processes which can be used for the analysis of genomic data with the help of dynamical system framework. The research area for such a field is huge, and thus, the primary focus has been applied to the development of network models with the help of biological big data collected. The applications already developed for the analysis of biological data can be divided into two groups, each consisting of genetic regulatory network and reconstruction of the network-dependent metabolism [47]. Different methods of analysis vary in terms of performance and have been based on combination of different approaches in order to produce optimized predictions.

One of the few developed fields in the genomic analytics is the reconstruction of gene network of regulation. Networking interference methodology can be categorized in to five distinct categories based on the primary model in each case: mutual information, Boolean regulatory network, regression, correlation and the others [48, 49]. Different methodology makes use of different parts of the same information to provide with the optimal solution.

Boolean regulatory networks can be considered one of the special cases for the use of discrete dynamic models in which the state of a set of nodes exists only in binary state [50, 51]. The state of a particular node is determined with the help of Boolean equations and the values of other connecting nodes. Use of Boolean network has been found to be the most helpful when the data is in small amount but is prone to produce high number of false positives which can only be reduced with prior knowledge about the data. Another challenge, which the Boolean network face, is the cost incurred when there is a higher number of nodes in the network. The number of global states rises exponentially with respect to the number of entities in the data. This can be overcome with the help of clustering technique on the data [52]. The clustering of the data would help in the breakdown of the data into smaller groups, which help in faster analysis. The main dynamics of gene regulatory network analytics can be understood with the help of ordinary differential equations (ODE). The process of reconstructing the gene regulatory network in a genome-scale system is to be same as a dynamic model is computationally intensive. This reduces the computational time of the system down to $O(n^2)$ which is much lower than the other processes which require $O(n^3)$ or even higher $O(n^2 \log n)$. In order to compute the regulatory network for a human genome with as much as 35,000 genes, there would be the exploration of around a billion connections in the network [53, 54].

The current trend to digitize medical systems and move to electronic patient records has changed in the medical and healthcare industry. The amount of clinical information and data accessible electronically increases perilously in respect to ramification, diversity and seasonableness, ensuing in what is acknowledged as big data. With essential needs and improved treatments, saving lives and reducing costs, big data has an unprecedented dimension of convenience and use cases, along with key cases in clinical decision support, health insurance, disease surveillance, population health management, side effect monitoring and more. Adaptation of treatment to events and diseases affects multiple organ systems 1, 2. Adopting big data technology in healthcare has various advantages and possibilities, but it also poses some obstacles and challenges. Indeed, many growth tendencies in health care, such as clinical manoeuvrability and wireless networking, medical data and information exchange, cloud computing, and more raise concerns about the security and privacy of sensitive information. In addition, healthcare organizations have found that bottom-up, technology-driven addressing to conclusive security and privacy necessities is not enough and neither sufficient to conservation organizations and patients. Proactive approaches and measures that all medical institutions must take into account for future security and privacy requirements to avert the invasion of delicate and sensitive data and information and other various types of security incidents.

6 Privacy Versus Security in Health Care

The patient's right and will to administer or adjust the acknowledgement of personal health and medical information can be defined as confidential in medicine (medical). Confidentiality can also be stated by the fair information code (FIP) as there must be a way or an approach to withholding data and information about a patient that has been used or contrived convenient for other benefits without the consent of the patient. Providing confidentiality requires patient acquiesce. Consent includes anyone who has access to a patient's specific records for legitimate purposes [55]. Security can be explained as physical and technical assessment or using health data for unauthorized use. This can be done to protect against public or illegal use of restricted data that violates the access control policy needs to be prevented in this scenario. There are security targets that must be marked according to the ISO-OSI standard recommended for structure data confidentiality, data integrity and access administration. These objectives must be appropriately implemented to ensure data security in the healthcare field for providing patient data confidentiality in order to enforce these security objectives or policies and therefore personal information related to various security policies [56, 57]. These security queries explain the tools or security methods required for implementing these security policies

6.1 Issues Related to Security and Privacy

The implementation of technologies like electronic patient record (EPR) system or electronic health record (EHR), remote patient regulating and monitoring adopting sensor network, and a hybrid combination of electronic record and sensor network can improve the quality of treatment [58]. In this section, you need to analyse some security and privacy issues:

6.1.1 Storage and Access to Medical Big Data

As we are aware of that medical big data records are a widely popular use of sensor networks as EPR and remote network monitoring, the records become exposed to hackers, malicious attacks and illegitimate access that compromises concealment and isolation [59]. There are two ways to the storage of increased risk data. Data can be stored in a central database or in a local database linked to each database within the network. Choosing a storage plan is critical to addressing security and privacy concerns. Medical data must be disclosed without the consent of the patient in order to provide the necessary treatment within the minimum time of an emergency. You should evaluate your access policy according to your situation. In order to use patient records in EPR, users are stated into two categories based on their permission to access the records [60].

Edit-read/write access (e.g. doctor, nurse, etc.)

Read-only access (e.g. health insurance provider)

6.1.2 Lack of Public Trust

To enable treatment, medical data must be exposed to a genuine authorized person. The healthcare department implements appropriate procedures and policies to limit the disclosure of records and to explain appropriate levels of access [61, 62]. To elevate the quality of your data, you need confidence. In information technology (IT), White provides a disagreement of trust between the subjects participating in the communication and IT solution.

6.1.3 Differences Between Various Security and Privacy Rules/Laws/ Legal Frameworks and Standards in the Current Technical Medical (Medical) Environment, There Are Many Security and Privacy Regulations/Laws/ Legal Frameworks and Standards

HIPAA (U.S. Health Insurance Mobility and Accountability Act) for security policy. Simultaneously, multiple of these state laws/laws are recognized for example. The Health Privacy Project, the health privacy and medical health information technology status of the Economic and Clinical Health Act (HiTech), which includes laws that conflict with separate laws, open the door to security threats [63].

For example, there are many international standards inherited from different regions. High-Level International Version 7 (HL7), International Classification of Disease (ICD) and many other things increase the likelihood of malicious attacks.

6.1.4 Warehousing, Big Data Mining and Web Mining Technologies

The mining and storage of these patterns of big data are done in a data warehouse to analyse and identify big data through relationships and dependencies in the form of patterns [64]. Web mining is a unified information created with the support of classic mining methods. Mining medical big data demands storing, managing and analysing strict methods and barriers to regulate security and privacy [65]. Proper oversight and/or assessment of mining technology to increase efficiency is a major challenge.

6.2 Solutions

Here, we will discuss some of the solutions discussed by researchers regarding medical (health care) big data security and privacy issues:

6.2.1 Role-Based Approach Control/Security

It is a unique updated essential model for access control. Approach to specified tasks is the ability of all users [66, 67]. Roles are assigned according to their privileges depending on what assets are arranged or can be provided in what amounts. The access control list (ACL) needs to be modified periodically.

6.2.2 The Implication of Cryptographic Encryption Techniques

Encryption is the method of converting a plain text message into something that can only be read by the communication entity muddled in the data transmission. Encryption levels can be done in both software and hardware. There are multiple types of encryption algorithms like Triple Encryption (Data Encryption Standard), Advanced Encryption Standard (AES), International Data Encryption Standard (IDEA), etc.

6.2.3 Authentication

For example, many authentication algorithms are used. Digital signatures, cryptographic mechanisms and hash functions of the sensor network can be used for authentication. The most widely used algorithms can be implemented in sensor networks built by Kansas State University.

6.2.4 Review Plan for Security and Privacy

Here, security policies are implemented according to the needs of the organization. It is necessary to implement self-control protection.

Data protection laws are much more important than ever, where healthcare institutions manage personal data and arrange secure data to protect and enforce data and legal responsibilities in co-relation to the processing of personal data and information. Various countries have various data privacy policies and laws. Data protection regulations and laws in some countries are listed in Table 3 along with their main features [68].

7 Limitation and Future Research

The use of Hadoop has helped in the achievement of an effective patient care system, which works with the help of survey data collected, from different classes of citizens. A secured network can be set up with the help of Hadoop, which can be used for the implementation of big data analytics in a Linux environment. The Linux environment itself manages the access control. With the help of the case studies taken for this research, it has helped in the understanding the process which the healthcare industry can implement in order to gain access to the vast amount of data of the industry and make use of it to gain in their business. However, like all study, this study has limitations. Healthcare industries lack behind on the use of proper IT adoption from other industries. This acts as the main reason for cases to be found. There are not many cases on the use of IT in healthcare industries. Other than the use of empirical-based big data analytics, there needs to be the use of more scientific study in the field. With the rise in the amount of unstructured data, there needs to be the use of proper analytical and decision support mechanism which needs to be implemented. Machine learning can be used to find the correct data from the large amount of gathered unstructured data. The future research can be conducted in order to find an analytical algorithm, which would be helpful in order to process unstructured data.

Table 3 Data protection laws in some countries

Country	Law	Features
USA	HIPAA Act Patient safety and quality improvement act (PSQIA) HITECH Act	It is necessary to establish national standards for electronic medical transactions. It provides privacy rights to individuals between the ages of 12 and 18. Signed disclosure from affected people before providing information about health care provided to anybody, including parents. Patient safety work products must not be exposed to. 27. Individuals who violate confidentiality supplies are subject to civil penalties. Protect the security and privacy of electronic health information
EU	Data protection directive	It protects the basic rights and freedoms of people, especially the right to privacy in relation to the processing of personal information and data
CANADA	Personal information protection and electronic documents act ('PIPEDA')	Individuals have the right to know why personal information is collected or used, so organizations must protect this information reasonably and securely
UK	Data protection act (DPA)	It paves the way for individuals to regulate information related to themselves. Personal data will not be conveyed to any country or territory outside the European Economic Area except for the country or territory guarantees a competent level of protection for the rights and freedoms of the data subject
MOROCCO	The 09-08 act, dated on 18 February 2009	Protect your privacy by setting CNDP rights by restricting the implementation of personal and sensitive data employing a data controller in all data processing operations
RUSSIA	Russian federal law on personal data	Data operators are required to take all adequate organizational and technical steps to protect personal data from unauthorized or accidental infiltration

(continued)

Table 3 (continued)

Country	Law	Features
INDIA	IT Act and IT (Amendment) Ac	Use equitable security practices for delicate personal data or information. It delivers allowance to anyone troubled by unfair loss or unfair gain. We impose imprisonment and/or fines on those who take unreasonable losses or undue gains by exposing the personal data or information of others while delivering services in accordance with the terms of the legal contract
BRAZIL	Constitution	People's intimacy, privacy, honour and image are inviolable, and the overwhelming rights of material or moral damage arising from their violations are guaranteed
	Data protection law (Law no. 22/11of 17 June)	Concerning the processing of sensitive data, collection and dealing with is only permitted if legal provisions are authorizing such dealings and former approval of the APD has been fetched

8 Conclusion

To conclude this study, it can be said that the cases studied suggest that the use of big data analytics in the healthcare industry can be helpful for potentially increasing the capability and business benefits of the industry. Strategies for the implementation of the big data analytics in healthcare organizations have also been identified and discussed. A focus has been put into three main areas: analysis of medical imaging, genomic data processing and physiological signal processing. The exponential rise in the medical images has forced IT specialists to implement better processing technique in a shorter timeframe. There is a steady development of various systems, which makes use of physiological signals in order to save lives. In the end, the main dimensions for the development of IT business values are processes, people and IT. However, the study focuses on the analytical processing side ignoring the people factor. It can indirectly found from the study that inclusion of analytical personnel in the industry is just as important as the use of analytical processing. Thus, future research should make use of the people factor in order to make the analysis.

References

1. Ker JI, Wang Y, Hajli MN, Song J, Ker CW (2014) Deploying lean in healthcare: evaluating information technology effectiveness in US hospital pharmacies. Int J Inf Manag 34(4):556–560

2. Jiang P, Winkley J, Zhao C, Munnoch R, Min G, Yang LT (2014) An intelligent in-formation forwarder for healthcare big data systems with distributed wearable sen-sors. IEEE Syst J PP(99):1–9
3. Raghupathi W, Raghupathi V (2014) Big data analytics in healthcare: promise and po-tential. Health Inf Sci Syst 2(1):3
4. Watson HJ (2014) Tutorial: big data analytics: concepts, technologies, and applications. Commun Assoc Inf Syst 34(1):1247–1268
5. Sharma R, Mithas S, Kankanhalli A (2014) Transforming decision-making processes: a research agenda for understanding the impact of business analytics on organisations. Eur J Inf Syst 23(4):433–441
6. Cox M, Ellsworth D (1997) Application-controlled demand paging for out-of-core visualization. In: Proceedings of the 8th IEEE conference on visualization. IEEE Computer Society Press, Los Alamitos, CA
7. Bryant RE, Katz RH, Lazowska ED (2008) Big-data computing: creating revolutionary breakthroughs in commerce, science, and society computing. In: Computing Research Initiatives for the 21st Century. Computing Research Association (Available at http://www.cra.org/ccc/files/docs/init/Big_Data.pdf)
8. Wang Y, Kung L, Byrd TA (2018) Big data analytics: Understanding its capabilities and potential benefits for healthcare organizations. Technol Forecast Soc Chang 126:3–13
9. Hurwitz J, Nugent A, Hapler F, Kaufman M (2013) Big data for dummies. Wiley, Hoboken, New Jersey
10. LaLalle S, Lesser E, Shockley R, Hopkins MS, Kruschwitz N (2011) Big data, analytics and the path from insights to value. MIT Sloan Manag Rev 52(2):21–31
11. Wang Y, Hajli N (2017) Exploring the path to big data analytics success in healthcare. J Bus Res 70:287–299
12. Shang S, Seddon PB (2002) Assessing and managing the benefits of enterprise systems: the business manager's perspective. Inf Syst J 12(4):271–299
13. Burnard P (1991) A method of analysing interview transcripts in qualitative research. Nurse Educ Today 11(6):461–466
14. Dey I (1993) Qualitative data analysis. A User-friendly Guide for Social Scientists, Routledge, London
15. Downe-Wamboldt B (1992) Content analysis: method, applications, and issues. Health Care Women Int 13(3):313–321. The business manager's perspective. Inf Syst J 12(4):271–299
16. Archenaa J, Anita EM (2015) A survey of big data analytics in healthcare and government. Procedia Comput Sci 50:408–413
17. Ren Y, Werner R, Pazzi N, Boukerche A (2010) Monitoring patients via a secure and mobile healthcare system. IEEE Wirel Commun 17(1):59–65
18. Elshazly H, Azar AT, El-korany A, Hassanien AE (2013) Hybrid system for lymphatic diseases diagnosis. In: Proceedings of the international conference on advances in computing, communications and informatics (ICACCI '13), IEEE, Mysore, India, pp 343–347
19. Dougherty G (2009) Digital image processing for medical applications. Cambridge University Press
20. Bernatowicz K, Keall P, Mishra P, Knopf A, Lomax A, Kipritidis J (2015) Quantifying the impact of respiratory-gated 4D CT acquisition on thoracic image quality: a digital phantom study. Med Phys 42(1):324–334
21. Liebeskind DS, Feldmann E (2015) Imaging of cerebrovascular disorders: precision medicine and the collaterome. Ann New York Acad Sci
22. Hussain T, Nguyen QT (2014) Molecular imaging for cancer diagnosis and surgery. Adv Drug Deliv Rev 66:90–100
23. Mustafa S, Mohammed B, Abbosh A Novel preprocessing techniques for accurate microwave imaging of human brain. IEEE Antennas Wireless Propag Lett 12
24. Tempany CMC, Jayender J, Kapur T et al (2015) Multimodal imaging for improved diagnosis and treatment of cancers. Cancer 121(6):817–827

25. Shvachko K, Kuang H, Radia S, Chansler R (2010) The Hadoop distributed file system. In: Proceedings of the IEEE 26th symposium on mass storage systems and technologies (MSST '10). IEEE, May 2010, pp 1–6
26. Sobhy D, El-Sonbaty Y, AbouElnasr M (2012) MedCloud: healthcare cloud computing system. In: Proceedings of the international conference for internet technology and secured transactions, IEEE, London, UK, December 2012, pp 161–166
27. Shackelford K (2014) System & method for delineation and quantification of fluid accumulation in efast trauma ultrasound images. US Patent Application 14/167,448
28. Chen W, Cockrell C, Ward KR, Najarian K (2010) Intracranial pressure level prediction in traumatic brain injury by extracting features from multiple sources and using machine learning methods. In: Proceedings of the IEEE international conference on bioinformatics and biomedicine (BIBM '10), IEEE, December 2010, pp 510–515
29. Ohno-Machado L, Bafna V, Boxwala AA et al (2012) iDASH: integrating data for analysis, anonymization, and sharing. J Am Med Inform Assoc 19(2):196–201
30. Rolim CO, Koch FL, Westphall CB, Werner J, Fracalossi A, Salvador GS (2010) A cloud computing solution for patient's data collection in health care institutions. In: Proceedings of the 2nd International Conference on eHealth, Telemedicine, and Social Medicine (ETELEMED '10). IEEE, February 2010, pp 95–99
31. Jun SW, Fleming KE, Adler M, Emer J (2012) ZIP-IO: architecture for application-specific compression of Big Data. In: Proceedings of the international conference on field-programmable technology (FPT '12), December 2012, pp 343–351
32. Belle A, Thiagarajan R, Soroushmehr SM, Navidi F, Beard DA, Najarian K (2015). Big data analytics in healthcare. BioMed research international (2015)
33. Hu P, Galvagno SM Jr, Sen A et al (2014) Identification of dynamic prehospital changes with continuous vital signs acquisition. Air Med J 33(1):27–33
34. Chen J, Dougherty E, Demir SS, Friedman CP, Li CS, Wong S (2005) Grand challenges for multimodal bio-medical systems. IEEE Circuits Syst Mag 5(2):46–52
35. McCullough JS, Casey M, Moscovice I, Prasad S (2010) The effect of health information technology on quality in U.S. hospitals. Health Aff 29(4):647–654
36. Kaur K, Rani R (2015) Managing data in healthcare information systems: many models, one solution. Computer 48(3):52–59
37. Prasad S, Sha MSN (2013) NextGen data persistence pattern in healthcare: polyglot persistence. In Proceedings of the 4th international conference on computing, communications and networking technologies (ICCCNT '13), July 2013, pp 1–8
38. Yu WD, Kollipara M, Penmetsa R, Elliadka S (2013) A distributed storage solution for cloud based e-Healthcare Information system. In: Proceedings of the IEEE 15th international conference on e-Health networking, applications and services (Healthcom '13). Lisbon, Portugal, October 2013, pp 476–480
39. Uzuner O, South BR, Shen S, DuVall SL (2011) 2010 i2b2/VA challenge on concepts, assertions, and relations in clinical text. J Am Med Inform Assoc 18(5):552–556
40. Athey BD, Braxenthaler M, Haas M, Guo Y (2013) tranS-MART: an open source and community-driven informatics and data sharing platform for clinical and translational research. AMIA Summits Transl Sci Proc 2013:6–8
41. Mishra S, Mishra BK, Tripathy HK, Dutta A (2020) Analysis of the role and scope of big data analytics with IoT in health care domain. In: Handbook of data science approaches for biomedical engineering. Academic Press, pp. 1–23
42. Mishra S, Tripathy HK, Mishra BK, Sahoo S (2018) Usage and analysis of big data in E-health domain. In: Big data management and the internet of things for improved health systems. IGI Global, pp 230–242
43. Mishra S, Tripathy HK, Mishra BK (2018) Implementation of biologically motivated optimisation approach for tumour categorisation. Int J Comput Aided Eng Technol 10(3):244–256
44. Seely JE, Bravi A, Herry C et al (2014) Do heart and respiratory rate variability improve prediction of extubation outcomes in critically ill patients?. Crit Care 18(2), article R65

45. Le Roux P, Menon DK, Citerio G et al (2014) Consensus summary statement of the international multidisciplinary consensus conference on multimodality monitoring in neurocritical care. Intens Care Med 40(9):1189–1209
46. Lander ES, Linton LM, Birren B et al (2001) Initial sequencing and analysis of the human genome. Nature 409(6822):860–921
47. Drmanac R, Sparks AB, Callow MJ et al (2010) Human genome sequencing using unchained base reads on self-assembling DNA nanoarrays. Science 327(5961):78–81
48. Andre F, Mardis E, Salm M, Soria JC, Siu LL, Swanton C (2014) Prioritizing targets for precision cancer medicine. Ann Oncol 25(12):2295–2303
49. Mishra S, Sahoo S, Mishra BK (2019) Neuro-fuzzy models and applications. In: Emerging trends and applications in cognitive computing. IGI Global, pp 78–98
50. Mishra S, Sahoo S, Mishra BK (2019) Addressing security issues and standards in Internet of things. In: Emerging trends and applications in cognitive computing. IGI Global, pp 224–257
51. Rath M, Mishra S (2020) Security approaches in machine learning for satellite communication. In: Machine learning and data mining in aerospace technology. Springer, Cham, pp 189–204
52. Rath M, Mishra S (2019) Advanced-level security in network and real-time applications using machine learning approaches. In: Machine learning and cognitive science applications in cyber security. IGI Global, pp 84–104
53. Khatri P, Draghici S, Ostermeier GC, Krawetz SA (2001) Profiling gene expression using Onto-Express. Genomics 79(2):266–270
54. Zeeberg BR, Feng W, Wang G et al (2003) GoMiner: a resource for biological interpretation of genomic and proteomic data. Genome Biol 4(4), article R28 (2003)
55. Bindea G, Mlecnik B, Hackl H et al (2009) Cluego: a cytoscape plug-in to decipher functionally grouped gene ontology and pathway annotation networks. Bioinformatics 25(8):1091–1093
56. Mishra S, Tripathy N, Mishra BK, Mahanty C (2019) Analysis of security issues in cloud environment. Secur Des Cloud, Iot, and Soc Network, pp 19–41
57. Mishra S, Mahanty C, Dash S, Mishra BK (2019) Implementation of BFS-NB hybrid model in intrusion detection system. In: Recent developments in machine learning and data analytics. Springer, Singapore, pp 167–175
58. Mishra S, Tripathy HK, Mallick PK, Bhoi AK, Barsocchi P (2020) EAGA-MLP—an enhanced and adaptive hybrid classification model for diabetes diagnosis. Sensors 20(14):4036
59. Bindea G, Galon J, Mlecnik B (2013) CluePediaCytoscape plugin: pathway insights using integrated experimental and in silico data. Bioinformatics 29(5):661–663
60. Mootha VK, Lindgren CM, Eriksson K-F et al (2003) PGC-1 -responsive genes involved in oxidative phosphorylation are coordinately downregulated in human diabetes. Nat Genet 34(3):267–273
61. Mishra S, Mallick PK, Jena L, Chae GS (2020) Optimization of skewed data using sampling-based preprocessing approach. Front Pub Health 8:274. https://doi.org/10.3389/fpubh.2020.00274
62. Dutta A, Misra C, Barik RK, Mishra S (2021) Enhancing mist assisted cloud computing toward secure and scalable architecture for smart healthcare. In: Hura G, Singh A, Siong Hoe L (eds) Advances in communication and computational technology. Lecture notes in electrical engineering, vol 668. Springer, Singapore. https://doi.org/10.1007/978-981-15-5341-7_116
63. Mohapatra SK, Nayak P, Mishra S, Bisoy SK (2019) Green computing: a step towards eco-friendly computing. In: Emerging trends and applications in cognitive computing. IGI Global, pp 124–149
64. Thiele NS, Fleming RMT et al (2013) A community-driven global reconstruction of human metabolism. Nat Biotechnol 31(5):419–425
65. Marbach D, Costello JC, Kuffner R et al (2012) Wisdom of crowds for robust gene network inference. Nature Meth 9(8):796–804
66. Mallick PK, Mishra S, Chae GS (2020) Digital media news categorization using Bernoulli document model for web content convergence. Pers Ubiquit Comput. https://doi.org/10.1007/s00779-020-01461-9

67. Sahoo S Mishra BP Jena N (2016) Building a new model for feature optimization in agricultural sectors. In: 2016 3rd international conference on computing for sustainable global development (INDIACom), New Delhi, pp 2337–2341 (2016)
68. Panda B, Mishra S, Mishra BKA Meta-model implementation with tabu search technique to determine the buying pattern of online customers. Ind J Sci Tech 9:S1

Chapter 9
An Analytical Perspective of Machine Learning in Cybersecurity

Rasika Kedia and Subandhu Agravanshi

1 Introduction

Ever since the dawn of intelligent beings, security, privacy and the sense of being safe in one's environment have been essential. The money and capital which drive start-ups, businesses and mammoth firms of all sectors are being handled by banks and finance firms and investment corporations with the objective of growth, better returns and safety. The finance sector today handles more money, data and sensitive information on clients, enterprises and consumers. Though these results in better insights, analytics and with the rise of data science help firms to grow and increase in capacity, it has also resulted in a rise of cybersecurity concerns. Since data is available in the digital format, it falls prey to hackers and the gray market. At the intersection of cybersecurity, finance sector, data analytics and machine learning, therefore, there is a growing need to ideate, innovate and implement strategies so as to preserve the data prone to attacks with the help of analytic tools and intelligent machines. The role of this paper is to implement machine learning to detect and segregate cybersecurity threats to data and helping firms tackle the same.

2 The Security Concerns

Via the three fundamental components of data protection, classification, trustworthiness and availability, cyberattacks can influence companies. Classification concerns

R. Kedia (✉) · S. Agravanshi
School of Computer Engineering, Kalinga Institute of Industrial Technology (KIIT), Deemed to be University, Bhubaneswar, Odisha, India
e-mail: 1805050@kiit.ac.in

S. Agravanshi
e-mail: 1805534@kiit.ac.in

P. K. Das et al. (eds.), *Privacy and Security Issues in Big Data*, Services and Business Process Reengineering, https://doi.org/10.1007/978-981-16-1007-3_9

occur as confidential data inside a corporation is revealed to others owing to which information penetrates. Problems of trustworthiness are correlated with framework abuse, similar to the scenario of misrepresentation. Finally, accessibility concerns are attributed to business disruptions in the long term. Although firms try their best to keep data safe, with new data being generated every second, it becomes really difficult to safeguard. Moreover with advancements rapidly around, the existing technologies often become obsolete, and in the buffer between security safeguards, hacks and security, attacks take place (Fig. 1).

Seamless data sharing of consumers and clients forms the backbone of the current finance market as it permits these firms in guiding and consultancy issues, thereby providing services, products and better prices. However, this makes the consumer data open and hardly a matter of personal space. This might result in sensitive information breaches as is the case with multiple confidentiality issues the banks often face. Confidentiality issues thereby lead to the clients dissatisfied with the firms which have failed to preserve their anonymity and hence have far-leading consequences. After client confidentiality, business disruptions due to data breaches are the next major concern which needs to be analyzed. Due to danger focus (Kopp et al. 2017) and the lack of substitutes on account of financial market infrastructures (FMIs), a business disruption of a financial sector system or a set of enormous financial pillars may have a tremendous impact. In the rare possibility, if an installment and settlement framework were disconnected during the day, market members will not be able to handle exchanges and raise dissolvability and liquidity hazards in this way. Correspondingly, on the off chance that one or several enormous banks are disturbed and incapable to handle exchanges, their partners would be dependent upon liquidity and dissolvability hazard.

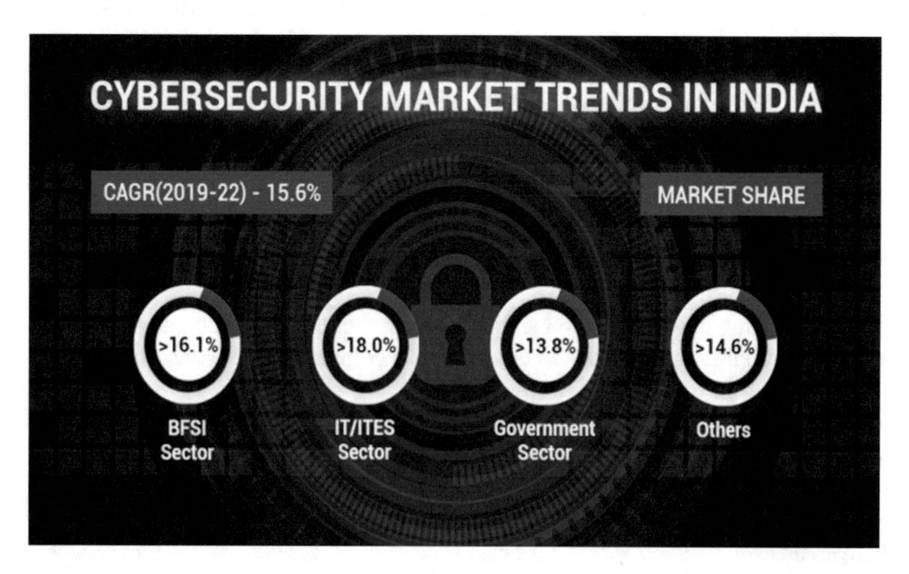

Fig. 1 Cybersecurity trending in market

Cyberattacks can also be utilized to sabotage clients' trust in an organization. For instance, on June 27, 2014, Bulgaria's biggest homegrown bank FIB encountered a contributor run, in the midst of elevated vulnerability because of the goal of another bank—following phishing messages demonstrating that FIB was encountering a liquidity deficiency. Stores surges on that day added up to 10% of the banks' total stores, and the bank needed to utilize a liquidity help plot given by the specialists. Once data breaches occur, confidentiality issues arise and business disruptions have taken place, the immense data is now open to attacks. This results in misuse of information for committing frauds. Frauds include using client data for fraudulent transactions, wherein 90% of these frauds result in the client losing money. As specialized, service-oriented firms have more data, high dependency on online transactions and technologies among other tech-driven support and lesser risk management and control over data flow, these firms are prone to leaving multiple entry points for hackers to get in and extract information. A number of these specialized firms also outsource their technology-based services to other third-party service providers. This is a serious privacy concern, not only because a lot of this data is being shared with a third-party firm without the clients/consumers being asked for consent or being made aware of, but also because these external service providers come with their own risks, and if they fall prey to a security breach or attack, they would disrupt multiple agencies at once.

Dealing with the digital characters of individuals and endeavors is a significant challenge for fintech organizations as associations mean to give a coordinated omnichannel experience to clients by broadening a large group of banking, wealth the executives and installment administrations in a consistent manner. Progressively, gadgets, for example, cell phones outfitted with biometric sensors (e.g., unique mark scanners), are being utilized to give confirmation and approval administrations. The utilization of cell phones as verification gadgets, using biometrics, once passwords (OTPs) and code-creating applications (e.g., Google Authenticator), has diminished the dependence on conventional confirmation systems, for example, passwords and PINs. While digital personalities have gotten more secure at one level, given the pervasive idea of their utilization in the developing fintech world, cloning of these characters can prompt enhanced dangers. These are just various security challenges which finance firms and banks face. An overview of the current methodologies of tackling these problems might assist us in further analyzing where machine learning steps in to the aid (Fig. 2).

3 Existing Methods of Security

Under confidentiality and data ownership, one manner by which organizations could beat the potential danger of prosecution (over leak or misuse of data) is by authorizing instruments which safely discard client information once he/she de-buys in from the utilization of fintech administrations. Further, overseeing client admittance

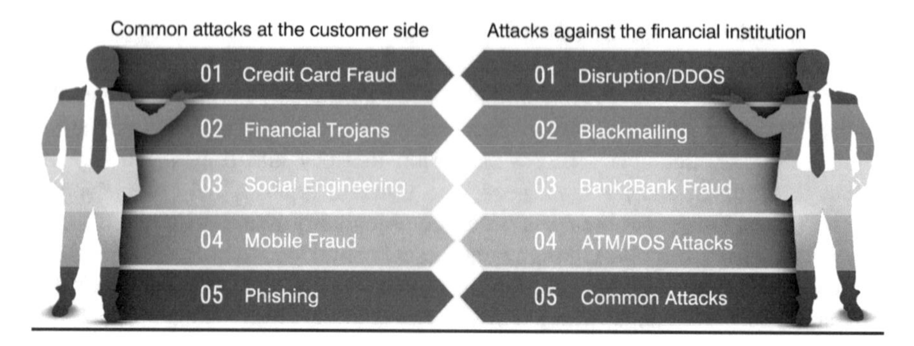

Fig. 2 Common vulnerabilities in cyberspace

to arrangements and administrations turns out to be progressively astounding. Cybersecurity ideas like information naming, specific information sharing and personality mindful information sharing hold potential answers for this issue. Versatile validation or danger-based confirmation, which analyzes client conduct in settling on choices on giving access, could be a countermeasure to address the danger of abuse of digital personalities. Installing security as a component of the initial plan stage by recognizing business use cases and creating danger models and related controls is one potential method to guarantee the advancement of secure innovations. This would significantly surrender the beginning idea of the fintech world and thus the danger of uncertain coding works on being received in pockets.

4 Role of Machine Learning in Dealing with Security Breaches

The effect of machine learning on the banking, financial services and insurance (BFSI) sector is unrivaled in light of the fact that ML-based calculations anticipate more exact outcomes when taken care of with more information. Fortuitously, the BFSI segment accommodates petabytes of information on the client, exchanges, solicitations, cash moves and so forth to help these calculation-based models learn and improve ceaselessly. SupTech is the utilization of ML-empowered advancements by open segment administrative establishments to create applications that upgrade productivity and viability of oversight and surveillance. These ML-based methods are additionally utilized by monetary administrations to recognize dangers, for example, loaning cheats, ATM hacks, illegal tax avoidance and digital assaults. A portion of these methods is likewise used to create models that help decide shrouded designs in unstructured informational collections that are difficult to follow utilizing regular measurable systems. In the whole BFSI division, ML models help with distinguishing and featuring fake cases for additional human examination and help guarantee a goal dynamic cycle to dodge human predisposition. Not just this, these organizations use

ML-based answers for screen movement levels of clients and prize great conduct of the client by offering them limits or suitable kinds of approaches. With such a comprehensive client view, insurance agencies can all the more likely oversee hazards as well. As fraudsters continually change their systems to dodge recognition, ML methods can be utilized to spot moving examples, prompting improved location rates while lessening countless false positives. Machine learning essentially dissects the information from an earlier time and assesses the utilization cases for what is to come. The calculations dependent on AI and machine learning are taken care of by the information-driven experiences during that time of information relating to the action logs. In light of this, AI calculations anticipate future events and client conduct and recommend proactive measures likewise. Machine learning-based solutions likewise empower associations to distinguish expanding measures of classified information that require insurance. For instance, Symantec uses vector machine learning (VML) innovation in identifying delicate data from unstructured information. Through preparing, this methodology can improve the precision and unwavering quality of discovering touchy data persistently. Hart et al. introduced machine learning-based text characterization calculations to naturally recognize touchy or non-sensitive undertaking archives. Alneyadi et al. utilized factual investigation strategies to recognize private information semantics in developed information that seems fluffy in nature or has different varieties. As per PwC's "Artificial Intelligence in India—promotion or reality" report, leaders in the banking, financial services and insurance industry (BFSI) alluded to machine learning, robotized data analysis and robotics as the three fundamental AI-energized arrangements with the biggest impact on their business. This is according to the business consideration on risk, the board, customer assistance and process mechanization (Fig. 3).

5 Challenges ML Faces in Cybersecurity

Machine learning feeds, runs and progresses off the data that is generated all around. In the finance sector, there is just so much data which needs to be protected and preserved, but as previously mentioned, due to multiple reasons, this does not seem to be an easy task. If threats to security and data can be detected, segregated and conveyed to firms beforehand, or at least a few determining factors can be chalked out which predict where the threats are arising from, it will be easier to implement the appropriate algorithms. However, big data analytics, machine learning and cybersecurity all come with their own share of challenges and issues which must be previously understood and overcome before work on the intersection can begin. We must first analyze in detail these fields. Disruption issues consist of hackers getting into businesses and servers, thereby causing Internet leakages via use of ransomware and malware which heavily disrupts and hijacks the Internet of things.

For finance firms, their credibility and accountability to their clients and consumers are of utmost importance as that creates better opportunities for networks and businesses; unfortunately, hackers and bots causing misleading information

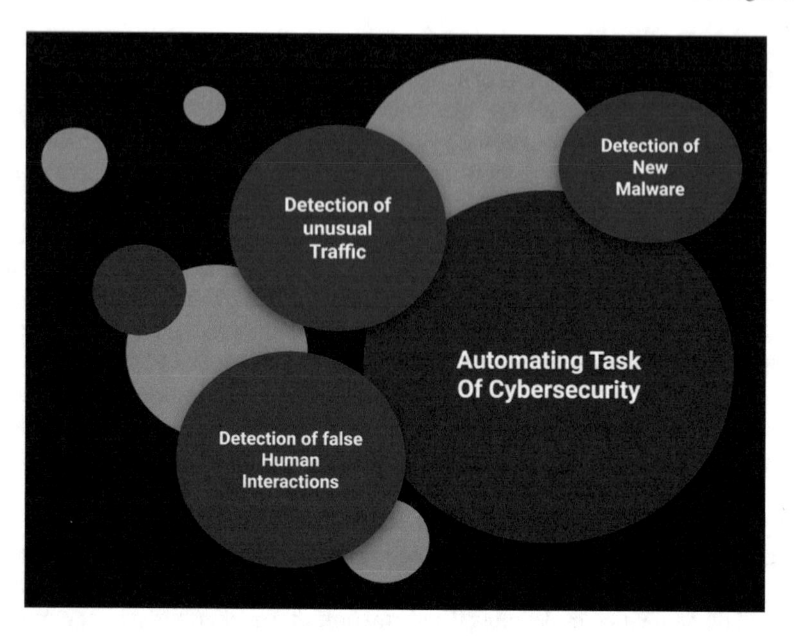

Fig. 3 Common uses of automating cybersecurity tasks

might end up causing vulnerability of company data causing compromise of trust and integrity of information. To research the online protection issues in IT frameworks, one, above all else, needs admittance to proper datasets. Without pertinent datasets, we just cannot assess the security dangers and dangers by any stretch of the imagination. Consequently, the significant obstruction to exploring network protection issues is the absence of admittance to suitable informational indexes for danger and danger assessment. All in all, organizations do not in general share their information, and network safety information is much more touchy than expected. There are numerous significant digital protection informational indexes like Microsoft's malware informational collection, EndGame's Ember malware properties dataset and Los Alamos' traffic dataset. Notwithstanding, we feel that there are no reasonable datasets that will empower scholarly specialists to adapt to the issues and difficulties that are experienced while executing ML calculations. One of the most widely recognized difficulties while actualizing AI is the issue of imbalanced datasets. When the proportions between the dominant parts of the sets and the minority parts of the sets are huge, a data set is then imbalanced. Where ordinarily a proportion of 1 to 10 is viewed as irregularity in the machine learning community, malignant training examples in cybersecurity datasets are normally very uncommon and may generally cause an imbalance of 1 to 10,000. At the point, when the proportion between the sets is enormous, yet there are sufficient examples in each set, we utilize the term relative imbalance. In the more extreme case, wherein there are insufficient examples in the minority set, this is known as absolute imbalance. Another challenge to be overcome while executing machine learning calculations is the domain adaptation. A situation wherein the distribution of the test dataset upon which a model is assessed varies from the distribution of the train dataset that was utilized to construct the model is known as domain adaptation. In every one of its areas, the cyberworld varies. There are different risks, networking activities and entities between parties operating on various scales, domains, geo-locations and cultures. Traditional ML algorithms usually adjust ineffectively to domain shifts. Thus, an essential challenge is adjusting to an effective defense mechanism from one domain onto the next. Most digital endeavors are fundamentally regulated learning undertakings. Given a component or activity, we should pick whether it is malignant or benevolent. Unfortunately, normally, we miss the mark on the names which are required for managed learning. Manual naming is limited in extension and unique. Along these lines, most names are beginning from heuristics. Rather than having mistakes, one may eventually wind up endeavoring to demonstrate the heuristics he began with (Fig. 4).

While a security expert adapts effectively to the majority of the difficulties and sends a high-performing framework to the creation condition, the framework endures a serious degradation in execution after a short period of time. Despite the fact that the model did not transform, the world around changed. Cybersecurity is a quickly advancing field. Models and perceptions are probably going to become outdated rapidly. Volumes change because technological advancement, new conventions and domains show up, and malevolent action likewise has patterns and designs. Concept drift is the transition in the source of the analyzed entity because of a transition in time. Approaches to follow concept drift and extraordinary changes in the domain should be created to keep away from continuously beginning from scratch.

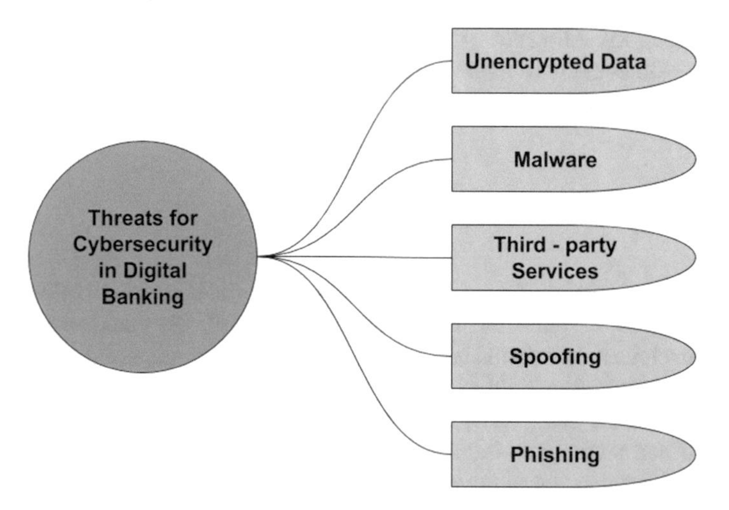

Fig. 4 Common cybersecurity threats in the BFSI domain

6 Demonstration of an Intelligent Cybersecurity Model

The below figure depicts an overview of the framework that involves various layers of processing, ranging from raw data to services. In the following, we have explained the working procedure of the framework in brief.

6.1 Security Data Collection

Gathering significant data on cybersecurity is an essential step, which connects the issues in cyberinfrastructure to the data-driven solutions in the above diagram. This is because cyberdata can form the foundation for determining the real attributes of the model that impact the performance of the model. Cyberdata's accuracy and reliability determine the workability and viability of solving the cybersecurity concern according to our target. The problem, however, is as to how we can collect usable and relevant data to create data-driven security models.

More often than not, the collection of data from multiple sources is based on a specific project and issue of an organization. Sources can be categorized into a variety of groups like host, network and hybrid. Under the host category, data is collected from the host machines, where the sources are database access logs, email logs, operating system logs, application logs, etc. Under the network category, various types of security data like network traffic data, firewall logs, packet data, honeypot data, IDS logs, etc., can be used by the security systems to provide the targeted security services. Collecting data from both the host machines and network is known as hybrid category. Altogether, database operation, network behavior, user activity and application behavior may be possible sources in the data collection layer when gathering cybersecurity data (Fig. 5).

6.2 Security Data Preparation

This layer is responsible for preparation of the raw data, after its collection from different sources, to build the model by implementing the necessary processes. However, irrelevant data should be removed since all the data captured by the network sniffer is not required to build the model. The accuracy of the model is proportional to the quality of the data. Thus, we require high-quality data for a higher accuracy in a data-driven model, i.e., training the model on a dataset and letting it figure out the function that gives us an output corresponding to a given input. The presence of missing/corrupted values, noisy data, or attributes of different scales and forms(like continuous, discrete or symbolic) requires pre-processing techniques such as data cleaning, handling of missing/corrupted values and converting the data to target type besides a reasonable understanding of the data, its types and the permissible

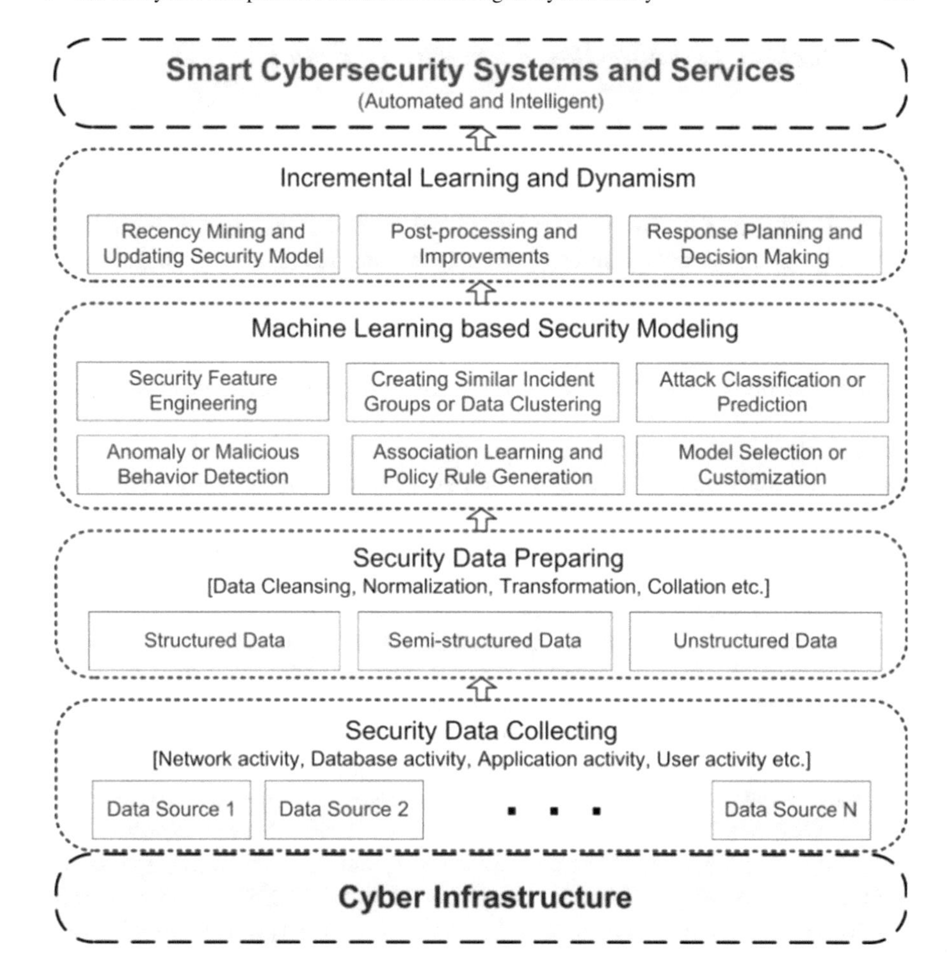

Fig. 5 Intelligent ML-based cybersecurity model

operations. Besides, transformation, collation or normalization is useful in arranging the data in a structured manner. Since both quantity and quality of data dictate the viability of addressing the cybersecurity issue, efficient management of data, their pre-processing and their representation may play a vital role in building a comprehensive data-driven security model for smart services.

6.3 Machine Learning-Based Security Modeling

It is one of the most crucial parts in gaining information and data knowledge via the implementation of network security data science, particularly machine learning-based modeling, since ML algorithms have the power to dramatically shift the landscape of cybersecurity. The attributes and features of the security data along with their patterns are of critical interest to be found and interpreted to obtain useful insights. In order to meet these goals, a comprehensive knowledge about data and ML-based frameworks leveraging a wide range of security data will be helpful. As a result, multiple machine learning activities may be used in this layer based on the solution viewpoint. One of them is feature engineering which is largely responsible for converting unprocessed data to insightful characteristics that successfully address the fundamental cybersecurity challenges for the data-driven models. As a consequence, multiple data processing activities like selection of features, by considering a subgroup of accessible security features on the basis of their associations or significance in modeling, or normalization and transformation, or creation and extraction of features by gaining new key components, can be included in this module, based on the characteristics of the data. Another critical module is security data clustering, which unveils concealed trends and patterns across large collections of the security data to determine where cyberthreats occur. Conventionally, it involves grouping security data with common features that can be used to address a variety of cybersecurity issues, like anomaly identification, policy breaches, etc. In the domain of cybersecurity, attack prediction or classification is considered to be among the most important modules which is responsible for constructing a predictive model for classifying threats or attacks and determining the future of a specific cybersecurity issue. The policy rule generation and association learning module will contribute to developing a skilled security system that contains a set of IF–THEN rules that identify threats. As an outcome, in a policy rule generation issue for a protocol-based access control framework, association learning might be used to discover links or connections within a collection of useful features in a cybersecurity dataset. The module model customization or selection is in charge of deciding whether to use the current ML model or whether to modify it. Data analysis and modeling on the basis of conventional ML or DL approaches might produce reliable findings in some cases in the field of cybersecurity. However, in terms of efficacy and productivity or other success metrics, taking into account computational time complexity, generalization ability and, most significantly, the algorithm's influence on the model prediction accuracy, ML models need to be tailored according to a particular security concern. In addition, customizing the appropriate data and techniques could boost the efficiency of the resulting security model making it even more applicable in the field of cybersecurity. The above-mentioned modules can function independently and collectively based on the security issues to be addressed.

6.4 Incremental Learning and Dynamism

The layer is associated with the validation of the resulting model by adding further understanding as per required. This may be accomplished by additional processing of various modules. The post-processing and improvements module might have an important part in simplifying the insights gained on the basis of specific needs by leveraging domain-specific knowledge. Since the machine learning approach-centered attack classification or prediction models depend heavily on training samples, it can barely be extended to other datasets that may be important for a few purposes. This module is accountable for using domain knowledge in the context of ontology or taxonomy to enhance attack association in cybersecurity applications in order to overcome those limitations.

A further important module recency mining and updating security model is accountable for updating the model according to the present-day needs by collecting the new data-driven security trends for increased efficiency. The previous layer's derived information is rooted on the original static dataset, taking into account the prevailing dataset trends. Nonetheless, owing to incremental security data with recent developments, such information cannot be promised higher precision in a variety of situations. Such incremental data might involve specific trends in some circumstances that may interfere with prior knowledge. Therefore, the RecencyMiner theory on the retrieval of emerging patterns and incremental data might be more powerful than the old trends that exist because new technology developments and techniques are more probable to be advantageous for detecting cyberbreaches or assaults than the earlier ones. Finally, in order to provide intelligent and automated services, the response planning and decision-making module is accountable for making judgments on the basis of the observations derived and taking the appropriate steps to avoid cyberthreats from taking place. The requirements may be different owing to the desired specifications of the security issue.

Overall, this framework is a generalized representation that can presumably be used to uncover vital information from security data, to develop smart cybersecurity solutions, to solve complicated security problems such as malware detection, access control management, denial-of-service attacks, detection of irregularities and theft in the field cybersecurity data science.

7 Common Machine Learning Algorithms that Are Used in Cybersecurity

Machine learning is normally regarded as part of "Artificial Intelligence," and is firmly identified with analytical knowledge, data processing and analytics, data science, especially zeroing in on letting machines learn from the information provided. Consequently, machine learning models usually include a set of protocols, strategies or complicated "transfer functions" which can be used to find intriguing

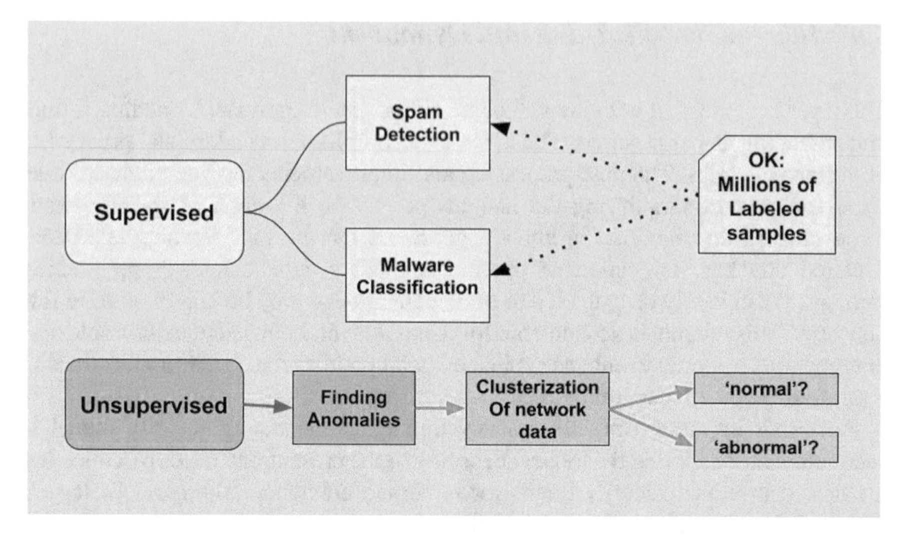

Fig. 6 A few use cases of ML algorithms in cybersecurity

information patterns, or to interpret or foresee actions, that could play an important role in the region of cybersecurity. In the accompanying, we talk about various strategies that can be utilized to tackle AI undertakings and how they are identified with cybersecurity assignments (Fig. 6).

7.1 Supervised Learning

It is implemented where certain objectives are described to be met from a particular course of action of information sources, indicating task-driven procedure. The most prominent supervised learning mechanisms in the AI domain are the regression and classification techniques. Such strategies are common in classifying or forecasting the future for a specific safety concern. As an example, classification methods may be applied in the cybersecurity domain to anticipate denial-of-service (DOS) attacks (yes, no) or to classify various groups of network attacks, such as spoofing and scanning. Popular classification strategies are adaptive boosting, ZeroR, logistic regression, Naives Bayes, KNN, decision tree, SVM and OneR. Additionally, IntruDTree and BehavDT classification strategies have lately been suggested by Sarker et al. which are able to efficiently create a data-driven prediction model. Alternatively, regression methods are useful, to estimate the continual or numerical value, for example, overall phishing attacks over a given time or to estimate the parameters of the network packet. They may equally be utilized to address the underlying causes of cybercrime and other kinds of scam. SVM and linear regression are known to be the standard methods of regression. Key distinction among regression and classification is that the predicted value is numeric or continual in regression, while categorical or discontinuous is the predicted value for classification. Committee-based learning is an expansion of supervised learning where multiple typical methods are combined, such as random forest learning, which creates several decision trees to resolve a specific protection problem.

7.2 Unsupervised Learning

In unsupervised learning issues, the fundamental errand is to discover examples, structures or information in unlabeled information, i.e., information-driven methodology. In the region of network protection, digital assaults like malware remain covered up somehow or another, incorporate changing their conduct powerfully and self-sufficiently to evade recognition. Bunching strategies, a kind of unaided learning, can assist with revealing the concealed examples and structures from the datasets, to distinguish markers of such modern assaults. Additionally, in recognizing abnormalities, strategy infringement, identifying and wiping out uproarious occurrences in information, bunching strategies can be helpful. K-implies and K-medoids are the mainstream parceling grouping calculations, and single linkage or complete linkage is the notable progressive bunching calculations utilized in different application areas. In addition, a base up bunching approach proposed by Sarker et al. can likewise be utilized by considering the information qualities.

Moreover, extraction and selection of features could be ideal for studying specific security issues. As of late, Sarker et al. have proposed a methodology for choosing security highlights as per their significance score esteems. Also, principal segment examination, straight discriminant investigation, Pearson connection investigation or non-negative network factorization are the mainstream dimensionality decrease strategies to explain such issues. Affiliation rule learning is another model, where AI-based strategy rules can forestall digital assaults. In a specialist framework, the guidelines are normally physically characterized by an information engineer working in a joint effort with an area master. Affiliation rule learning unexpectedly is the disclosure of rules or connections among a lot of accessible security highlights or traits in a given dataset. To evaluate the quality of connections, relationship investigation can be utilized. Numerous affiliation rule mining calculations have been proposed in the zone of AI and information mining writing, for example, rationale-based, continuous example-based, tree-based and so on. As of late, Sarker et al. have proposed an affiliation rule learning approach considering non-repetitive age that can be utilized to find a lot of valuable security strategy rules. Additionally, algorithms such as Apriori, FP-Tree, and RARM, Apriori-TID and Apriori-Hybrid, AIS, and Eclat are skilled to tackle such issues by creating a lot of strategy rules in the realm of cybersecurity.

7.3 Neural Networks and Deep Learning

Deep learning is a piece of AI in the region of man-made reasoning, which is a computational model that is motivated by the natural neural organizations in the human cerebrum. Counterfeit neural network (ANN) is often utilized in deep learning, and the most well-known neural organization calculation is backpropagation. It performs learning on a multi-layer feed-forward neural organization which comprises an info layer, at least one shrouded layers and a yield layer. The principle distinction between deep learning and old style AI is its exhibition on the measure of security information increments. Commonly, deep learning calculations perform well when the information volumes are enormous, while AI calculations perform nearly better on little datasets. In our previous work, Sarker et al., we have shown the viability of these methodologies considering logical datasets. In any case, deep learning approaches emulate the human mind component to decipher huge measure of information or the perplexing information, for example, pictures, sounds and messages. As far as highlight extraction to assemble models, deep learning lessens the exertion of planning an element extractor for every issue than the old style AI strategies. Close to these attributes, deep learning normally sets aside a long effort to prepare a calculation than an AI calculation; notwithstanding, the test time is actually the inverse. In this way, deep learning depends more on superior machines with GPUs than old style AI calculations. The most well-known deep neural organization learning models incorporate multi-layer perceptron (MLP), convolutional neural organization (CNN), intermittent neural organization (RNN) or long momentary memory (LSTM) network. As of late, specialists utilize these deep learning methods for various purposes, for example, distinguishing network interruptions, malware traffic recognition and characterization, and so forth in the area of cybersecurity.

7.4 Other Learning Techniques

Semi-supervised learning can be depicted as a hybridization of supervised and unsupervised strategies examined above, as it deals with both the named and unlabeled information. In the territory of cybersecurity, it could be helpful, when it needs to name information consequently without human mediation, to improve the exhibition of cybersecurity models. Fortification methods are another kind of AI that portrays an operator by making its own learning encounters through collaborating straightforwardly with nature, i.e., condition-driven methodology, where the earth is normally detailed as a Markov choice cycle and take choice dependent on a prize capacity. Monte Carlo learning, Q-learning, deep Q-networks are the most widely recognized support learning calculations. For example, in an ongoing work, the creators present a methodology for distinguishing botnet traffic or malevolent digital exercises utilizing support learning joining with neural organization classifier. In another work, the creators examine about the use of deep fortification learning to interruption discovery for supervised issues, where they got the best outcomes for the deep Q-network calculation. With regards to cybersecurity, a few genetic algorithms could likewise be utilized to work out an equivalent class of learning issues.

Different sorts of AI methods talked about above can be helpful in the area of cybersecurity, to construct a successful security model. In a table underneath, we have summed up a few AI methods that are utilized to manufacture different sorts of security models for different purposes. Despite the fact that these models commonly speak to a learning-based security model, in this paper, we intend to zero in on an exhaustive cybersecurity information science model and applicable issues, so as to fabricate an information-driven savvy security framework. In the following segment, we feature a few examination issues and expected arrangements in the territory of cybersecurity information science.

8 Deep Learning Methods in Cybersecurity

8.1 Malware

The amount and assortment of assaults by malicious softwares are ceaselessly expanding, making it challenging to fight them utilizing conventional strategies. Deep learning gives a chance to manufacture generalizable models in order to self-sufficiently recognize and order malware. This will offer protection from little scope entertainers utilizing known malware and huge scope entertainers utilizing new sorts of malware to assault associations or people.

8.2 Domain Generation Algorithms and Botnet Detection

Domain generation algorithms are normally utilized malware devices that create enormous quantities of space names which can be utilized for hard to follow inter-changes with C2 workers. The enormous amount of differing area names makes it hard to impede noxious spaces like boycotting or sink-holing by utilizing standard strategies. DGAs are frequently utilized in an assortment of digital assaults, including spam crusades, robbery of individual information and execution of disseminated refusal of administration (DDoS) assaults. Botnets are normal apparatuses utilized for digital assaults and are right now recognized utilizing conduct recognition draws near (i.e., recognizing foundational patterns of botnets' activities throughout their life cycle). In any case, rule-based social models have issues which muddle their utilization in light of the fact that these practices frequently happen over significant time frame scales.

8.3 Drive-by Download Attacks

Aggressors frequently abuse program weaknesses. By misusing defects in modules, an assailant can divert clients from ordinarily utilized sites to sites where endeavor code powers clients to download malware and run it. Such assaults are known as drive-by download assaults. To identify furthermore, forestall these assaults, an altered CNN called an occasion de-noising convolutional neural organization (EDCNN) was proposed by Shibahara et al. The EDCNN's fundamental purpose is the potential to decrease the repercussions of premalignant URLs which are also used inside undermined sites as well as diverted from traded-off sites. The researchers compared the effectiveness of two prevailing strategies with the EDCNN proposed by them and utilized a couple of attributes to improve the efficiency of the identification execution in their EDCNN.

8.4 File Type Identification

In general, people are not really that successful in detecting information, which after it has been encoded is being exfiltrated. At this challenge, signature-based strategies are equally ineffective. Thus, DL was implemented by Cox et al. by using a signal processing approach to distinguish and categorize types of files by utilizing DBNs. From the data, three different feature types are created. One is to take the Shannon entropy with a 50% crossway over a 256-byte window. Then, the Shannon entropy values are interpolated cubically onto a 256-point consistent grid, imparting a constant length. By treating the byte sequence like a signal and then converting it into frequency space, the second feature set is generated. A histogram of the bytes is the third feature. A four-layered DBN, pre-trained utilizing stacked de-noising autoencoders, was the classifier. The dataset consisted of 4500 files in total, 500 files of nine distinct types each, and 100 files from every type were segregated for evaluation. 97.44% was the total precision of classifying the nine distinct file forms.

8.5 Organization Traffic Identification

Utilizing stacked autoencoders joined with a sigmoid layer to execute grouping, Wang utilized DL to perform traffic type distinguishing proof. The dataset utilized by Wang was TCP stream information from an interior organization and the payload bytes of every meeting. Inside this dataset, there were 58 distinctive convention types; nonetheless, HTTP was prohibited in light of the fact that it is anything but difficult to distinguish and spoke to a vast dominant part of the information. A three-layer stacked autoencoder extracts highlights, and the highlights are then taken care of into a sigmoid layer that performs characterization. On the 25 most normal residual conventions, barring HTTP, this organization had an exactness somewhere in the range of 91.74% and 100% and a review somewhere in the range of 90.9% and 100%, contingent upon the convention type.

8.6 SPAM Identification

One of the early researches in utilizing DL for identifying spam emails was done by Likas and Tzortzis. Based on common terms found in emails, they extracted features and used a DBN with three secret layers constructed from RBM units. The precision of the DBN was greater, by a slight number, than the precision of the SVM. The DBN reached 99.45%, 97.5% and 97.43% accuracies, while the SVM had 99.24%, 97.32% and 96.92% accuracies.

8.7 Border Gateway Protocol Anomaly Detection

The Border Gateway Protocol (BGP) is a Web convention which considers the trading of directing as well as reachability data among independent frameworks. This capacity is basic to the working of the Web, and misuse of BGP faults may cause DDoS assaults, rerouting, sniffing, burglary of organization geography information and so forth. It is consequently basic to recognize odd BGP occasions continuously to moderate any expected harms. Cheng et al. [169] extricated 33 detail highlights and utilized an LSTM, joined with calculated relapse to distinguish a typical BGP traffic with 99.5% precision, a sensational boost on non-DL strategies.

8.8 User Authentication

Shi et al. had the option to utilize WiFi signals produced by IoT gadgets to distinguish human conduct and physiological highlights dependent on their everyday action designs utilizing an autoencoder. The human practices that they investigated included regular everyday movement, for example, strolling, and fixed practices. By extricating channel state data (CSI) estimations from WiFi flags, an autoencoder could distinguish every individual client and allocate a "unique mark" that would embody their practices. The three-layer autoencoder was planned with the end goal that each layer played out an assignment. The principal layer performs action partition, where it isolates out various exercises. The subsequent layer had the option to characterize activities. The third layer performed singular distinguishing proof. At that point, a SVM layer performed parody identification. They accomplished 91% confirmation exactness with few subjects (Table 1).

Table 1 List of machine learning algorithms used in cybersecurity

Technique	Purpose	Reference
SVM	To classify various attacks such as DoS, Probe, U2R and R2L	Kotpalliwar et al. [1]
SVM	DDoS detection and analysis in SDN-based environment	Kokila et al. [2]
SVM-PSO	To build intrusion detection system	Saxena et al. [3]
FCM clustering and ANN	To build network intrusion detection system	Chandrasekhar et al. [4]
KNN	To reduce the false alarm rate	Meng et al. [5]
K-means and KNN	To build intrusion detection system	Sharifi et al. [6]
Naive Bayes	To build an intrusion detection system for multi-class classification	Koc et al. [7]
Decision tree and KNN	Anomaly intrusion detection system	Balogun et al. [8]
Genetic algorithm and decision tree	To solve the problem of small disjunct in the decision tree-based intrusion detection system	Azad et al. [9]
Decision tree and ANN	To measure the performance of intrusion detection system	Jo et al. [10]
Random forests	To build network intrusion detection systems	Zhang et al. [11]
Association rule	To build network intrusion detection systems	Tajbakhsh et al. [12, 13]
Behavior rule	To build intrusion detection system for safety critical medical cyberphysical systems	Mitchell et al. [14, 15]
Genetic algorithm	For prevention of cyberterrorism through dynamic and evolving intrusion detection	Hansen et al. [16]
Deep learning convolutional	Malware traffic classification system	Kolosnjaji et al. [17]
BFS-NB framework	Intrusion detection model	Mishra et al. [18]
Mobile computing model with enhanced security	Secured cloud technologies	Mishra et al. [19]
MIST technology	Cloud computing-assisted methods	Arijit et al. [20]
Neural networks	Advanced level secured model	Rath et al. [21]

9 Conclusion

Machine learning is about developing patterns and manipulating those patterns with algorithms. In order to develop patterns, you need a lot of rich data from everywhere because the data needs to represent as many potential outcomes from as many potential scenarios as possible. With machine learning, cybersecurity systems can analyze patterns and learn from them to help prevent similar attacks and respond to changing behavior. It can help cybersecurity teams be more proactive in preventing threats and responding to active attacks in real time. It can reduce the amount of time spent on routine tasks and enable organizations to use their resources more strategically. Machine learning can make cybersecurity simpler, more proactive, less expensive and far more effective.

References

1. Kotpalliwar MV, Wajgi R (2015) Classification of attacks using support vector machine (svm) on kddcup'99 ids database. In: 2015 Fifth international conference on communication systems and network technologies. IEEE, pp 987–990
2. Kokila R, Selvi ST, Govindarajan K (2014) Ddos detection and analysis in sdn-based environment using support vector machine classifier. In: 2014 sixth international conference on advanced computing (ICoAC). IEEE, pp 205–210
3. Saxena H, Richariya V (2014) Intrusion detection in kdd99 dataset using svm-pso and feature reduction with information gain. Int J Comput Appl 98:6
4. Chandrasekhar A, Raghuveer K (2014) Confederation of fcm clustering, ann and svm techniques to implement hybrid nids using corrected kdd cup 99 dataset. In: 2014 international conference on communication and signal processing. IEEE, pp 672–676
5. Meng W, Li W, Kwok L-F (2015) Design of intelligent knn-based alarm filter using knowledge-based alert verification in intrusion detection. Secur Commun Netw 8(18):3883–3895
6. Sharifi AM, Amirgholipour SK, Pourebrahimi A (2015) Intrusion detection based on joint of k-means and knn. J Converg Inform Technol 10(5):42
7. Koc L, Mazzuchi TA, Sarkani S (2012) A network intrusion detection system based on a hidden naïve bayes multiclass classifier. Exp Syst Appl 39(18):13492–13500
8. Balogun AO, Jimoh RG (2015) Anomaly intrusion detection using an hybrid of decision tree and k-nearest neighbor
9. Azad C, Jha VK (2015) Genetic algorithm to solve the problem of small disjunct in the decision tree based intrusion detection system. Int J Comput Netw Inform Secur 7(8):56
10. Jo S, Sung H, Ahn B (2015) A comparative study on the performance of intrusion detection using decision tree and artificial neural network models. J Korea Soc Dig Indus Inform Manag 11(4):33–45
11. Zhan J, Zulkernine M, Haque A (2008) Random-forests-based network intrusion detection systems. IEEE Trans Syst Man Cybern C 38(5):649–659
12. Tajbakhsh A, Rahmati M, Mirzaei A (2009) Intrusion detection using fuzzy association rules. Appl Soft Comput 9(2):462–469
13. Mishra S, Tripathy HK, Mallick PK, Bhoi AK, Barsocchi P (2020) EAGA-MLP—an enhanced and adaptive hybrid classification model for diabetes diagnosis. Sensors 20(14):4036
14. Mishra S, Mallick PK, Jena L, Chae GS (2020) Optimization of Skewed data using sampling-based preprocessing approach. Front Public Health 8:274. https://doi.org/10.3389/fpubh.2020.00274
15. Mitchell R, Chen R (2014) Behavior rule specification-based intrusion detection for safety critical medical cyber physical systems. IEEE Trans Depend Secure Comput 12(1):16–30
16. Hansen JV, Lowry PB, Meservy RD, McDonald DM (2007) Genetic programming for prevention of cyberterrorism through dynamic and evolving intrusion detection. Decis Supp Syst 43(4):1362–1374

17. Kolosnjaji B, Zarras A, Webster G, Eckert C (2016) Deep learning for classification of malware system call sequences. In: Australasian joint conference on artificial intelligence. Springer, New York, pp 137–149
18. Mishra S, Mahanty C, Dash S, Mishra BK (2019) Implementation of BFS-NB hybrid model in intrusion detection system. In: Recent developments in machine learning and data analytics. Springer, Singapore, pp 167–175
19. Mishra S, Mohapatra SK, Mishra BK, Sahoo S (2018) Analysis of mobile cloud computing: architecture, applications, challenges, and future perspectives. In: Applications of Security, Mobile, Analytic, and Cloud (SMAC) technologies for effective information processing and management. IGI Global, pp 81–104
20. Dutta A, Misra C, Barik RK, Mishra S (2021) Enhancing mist assisted cloud computing toward secure and scalable architecture for smart healthcare. In: Hura G, Singh A, Siong Hoe L (eds) Advances in communication and computational technology. Lecture Notes in Electrical Engineering, vol 668. Springer, Singapore. https://doi.org/10.1007/978-981-15-5341-7_116
21. Rath M, Mishra S (2019) Advanced-level security in network and real-time applications using machine learning approaches. In: Machine learning and cognitive science applications in cyber security. IGI Global, pp 84–104

Chapter 10
Business Intelligence Influenced Customer Relationship Management in Telecommunication Industry and Its Security Challenges

Lewlisa Saha, Hrudaya Kumar Tripathy, and Laxman Sahoo

1 Introduction

Big data analytics has become the driving of force of companies involving any kind of decision-making and pattern recognition. The advancing technologies for the database management, different social media platforms and mobile devices have enabled these companies to gather huge amount of data about their customers. All these platforms have been generating both structured demographic data and unstructured behavioral data [1]. When it comes down to the marketing aspect of the business intelligence, all these data generated contributes to the decision-making to gain insights on the customers. Traditionally marketing strategies mainly relied on the market surveys done to analyze the customer behavior. But with big data techniques, the key information can be collected by mining the data on different online platforms. Big data analysis can be divided into three groups: business intelligence, mobile analytics and social media analytics. Even though these three categories work in different way, they are still connected [2].

Business intelligence (BI) are strategies and technologies used to gain useful information from the data collected through different platforms by different firms. BI plays a very important role in developing the achievements of the organizations by recognizing new chances, detecting probable threats, finding new business observations, improving the decision-making process and many more [3, 4]. As of now, the results of BI are mainly derived from structured and internal data, which misses out on a lot of information from the external and unstructured data sources. Due to

L. Saha (✉) · H. K. Tripathy
Kalinga Institute of Industrial Technology, Bhubaneswar, Odisha, India

H. K. Tripathy
e-mail: hktripathyfcs@kiit.ac.in

L. Sahoo
Vellore Institute of Technology, Vellor, Tamil Nadu, India

P. K. Das et al. (eds.), *Privacy and Security Issues in Big Data*, Services and Business Process Reengineering, https://doi.org/10.1007/978-981-16-1007-3_10

its high potential in generating higher business value, big data has become the center of attention of the business world. As per a survey in 2009 by TDWI, 38% of the organizations involved in the survey have practiced advanced analytics and 85% of them stated that they would implement it in three years [5]. By implementing big data, enterprises can figure out the present status of the business and the constantly changing consumer behavior.

Customer relationship management is an aspect which comes along side business intelligence. Gaining customer satisfaction by abiding to their needs is the main target of any consumer-centric business, and this cannot be achieved without analyzing all the demographic and behavioral data available in the market [6]. This is where business intelligence comes into perspective to give a better insight by mining information from all the raw data available internal or external to the organization. CRM systems and BI when used in an integrated way offer a very inclusive approach for the customers leading to better customer profiling and a better customer relationship management. BI can spur growth in sales and revenue by detecting potential customers and converting them into long-term consumers, reducing the churn rate and rising sales among the existing consumers [7].

Telecommunication industry is one of the examples where BI influenced CRM tools are used extensively to make proper usage of billions of data generated by the companies. These data mostly includes call record details, customer profile records and network data. The major reason why telecommunication industry requires to implement such tools is not only limited to the size of the data, but also includes competitive market and high churn rate. To be able to address these issues and provide a better service to their customer is the major goal of every telecom company.

In this digital age, with the huge amount of data comes the biggest concern of security. Two of the most valued intangible resources of the telecommunication industry are their customers and the customers' data. Conventionally, it is normal for the telecommunication industry to provide protection to its customers' communication which has led to the trust between the service providers and the customers. This has built up to an indirect assumption that the data is protected by the telecom service providers. On one side, this assumption has helped the telecommunication sector with the opportunity to become a mediator which speaks for the security and privacy. As a consequence, they have been faced with the task of making sure that no harm comes to their subscribers. So, without attracting undue attention, it needs to provide protection or security.

As for the organization of this chapter, Sect. 2 discusses about the role of big data and big data analytics in business intelligence, Sect. 3 gives a detailed idea on how business intelligence and customer relationship management works together and how it influences the telecommunication industry, Sect. 4 elaborates on the security challenges faced by business organizations, particularly telecommunication sector, and some course of actions against these challenges and finally the conclusion in Sect. 5.

2 Role of Big Data in Business Intelligence

Big data deals with humongous amount of data which is far more complex and is continuously growing in number collected from different external and internal sources. Even though big data mainly concentrates on unstructured and unorganized data, it can also be used on semi-structured and structured data [8]. These sources can vary from different online platforms, mobile devices, business transactions, administrative data to medical records, sensors and geospatial devices and many more. Figure 1 shows the different domain areas of big data analytics. Since big data can be generated by both humans and machines, the data generation speed is very high. As stated by Dedić and Stanier [9], many definitions of big data are mainly volume focused. But big data is not only about the massiveness of the data, it includes many features which also must be taken into consideration. Many long-established definitions of big data are built on the basic elements of variety, velocity and volume. These three Vs represent data type variety, diverse data generation velocity and data volume. For data variety, big data can provide in-depth insights using both structured and unstructured data from transactions, store-based video, sales management data, consumer preferences and so on. In terms of data velocity, big data can allow real-time information sharing from local to national governments for better decision-making. As for data volume, an example would be Nielsen generating around 300,000 rows of real-time data per second and more than one billion records per month. Some modern definitions of big data include more features such as validity, veracity, variability, value, vagueness, vocabulary and venue. But some of these elements lack proper relevance and research.

Big data is being used everywhere these days, in the education system, heath care system, hotel management, telecommunication industry, biomedical research and so on. But irrespective of how huge and ever increasing the amount of data is, it is only of use when it can be properly analyzed. Big data analytics is considered as a business opportunity because of the information gained by analyzing all these

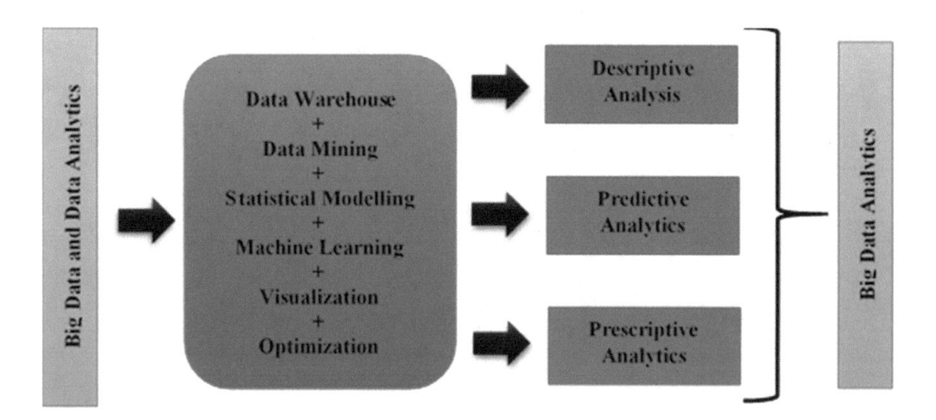

Fig. 1 Ontology of big data analytics

data. Since the main target of big data analytics is mining useful information and knowledge by analyzing the large data, implementing it in this fast-changing market can cause competitive advantage and improved responses from businesses [10].

BI is a technology-driven process which comprises data, products, strategies, technical architectures, applications, presentation and dissemination of business information. This helps the companies to get a better idea about their consumers' needs, which can help in building a better relationship with the customers and increase in revenue. It helps with the development of the organization by providing advantage over their competition in the market through achievement of positive information, maximizing profit and strategic decision-making. BI can also be used at operational level other than strategic and tactical purposes. It can obtain useful information and obscure knowledge from the regularly produced operational data. These materials can be of help to the business stakeholder in making predictions and calculations. Typically, BI would focus on extraction, transformation and loading (ETL), data warehousing and reporting which works for data manipulation, propagation and visualization. But the modernized BI also works on data exploration and visualization. The focus is also shifting toward interactive visualization which can find the cause and effect of event that has taken place. High market competition can lead to more research on BI and new upcoming trends [9]. Figure 2 shows how BI is related to big data.

Big data analytics is considered to be a part of BI since it helps with the decision-making in business using data, information and knowledge. Some of the leading BI tools are also used for big data analytics which suggests that they both support business decision-making by using some common tools. There are five main advantages of big data analytics: (1) Increase in perceptibility due to growth in accessibility of related data. (2) Improvement in performance because of accurate data. (3) Proper customer segmentation. (4) Automated algorithms that complement decision-making. (5) New business models, products and services.

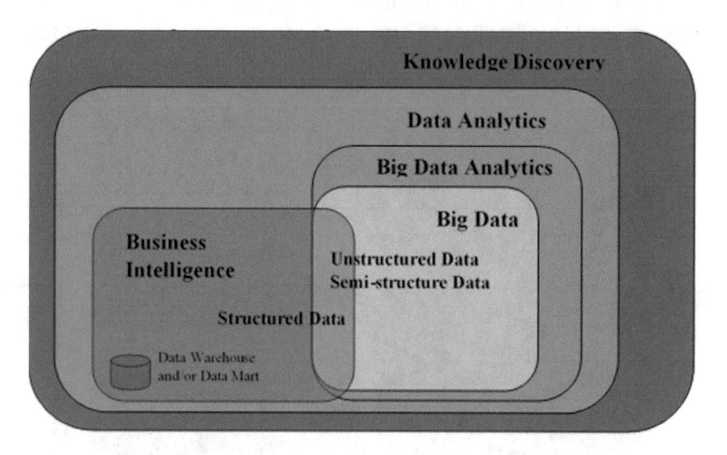

Fig. 2 Business intelligence versus big data [9]

Since BI deals with data and it is process driven, therefore, the flavor of BI can be incorporated in the telecom sector which will open up some new avenues which will help in decision-making and improving their relationship with their customers.

3 CRM and BI in Telecommunication Industry

Customer relationship management is a process used by almost every business sector for building a better relationship with their customers. It is used for a better customer interaction management using technology. The main target of CRM is to attract potential customers, understand their needs and satisfy them to earn their loyalty. CRM is used to retain customers for a long-lasting relationship by meeting their expectations which matches with the aim of any organization [11].

Due to the continuous surge in the amount of data, many companies use highly advanced computers with high storage capabilities to be able to implement big data analytics and to improve their performances which can be cost effective. There are both structured and unstructured data involved when it comes to business sectors and some of the sources for these data are retail databases, social marketing, customer activity records, enterprise data and logistics. These data sources help in establishing CRM strategies of higher quality level by being able to identify big data and its advantages. Even though big data analytics mainly reveals the type of data in the big data itself, some of the CRM strategies can also be done using big data analytics [12].

As discussed by Handzic et al. [12] in their work, developing a CRM strategy requires both business and customer strategy of an organization since the success of CRM is dependent on the interrelation of the two. First come the business strategies which contain the detailed analogy of the business and the target of the company concerning CRM and their competitions. The customer strategies are developed based on these business strategies. They need to work properly when put together. These strategies mainly involve analyzing the already existing and the potential customer group and proper customer segmentation for better product recommendation.

A major part of business decisions has the necessity to be supported by actual data which is the main benefit of CRM and BI. Both the processes provide the same advantages of getting a better understanding of the data to make better decisions, improve the relationship with the customers, predict and influence the performance and help with revenue growth. CRM will help in achieving a better BI, but BI can provide in-depth knowledge about the business acquired through multiple data sources.

Few of the impact that BI have over CRM are shown in Fig. 3 [13]:

The basic steps of implementing CRM and BI in the business to get better results are [14]:

1. Getting a better understanding of the customers to detect the key performance indicators (KPIs) that matches up with goals and objectives of the company.

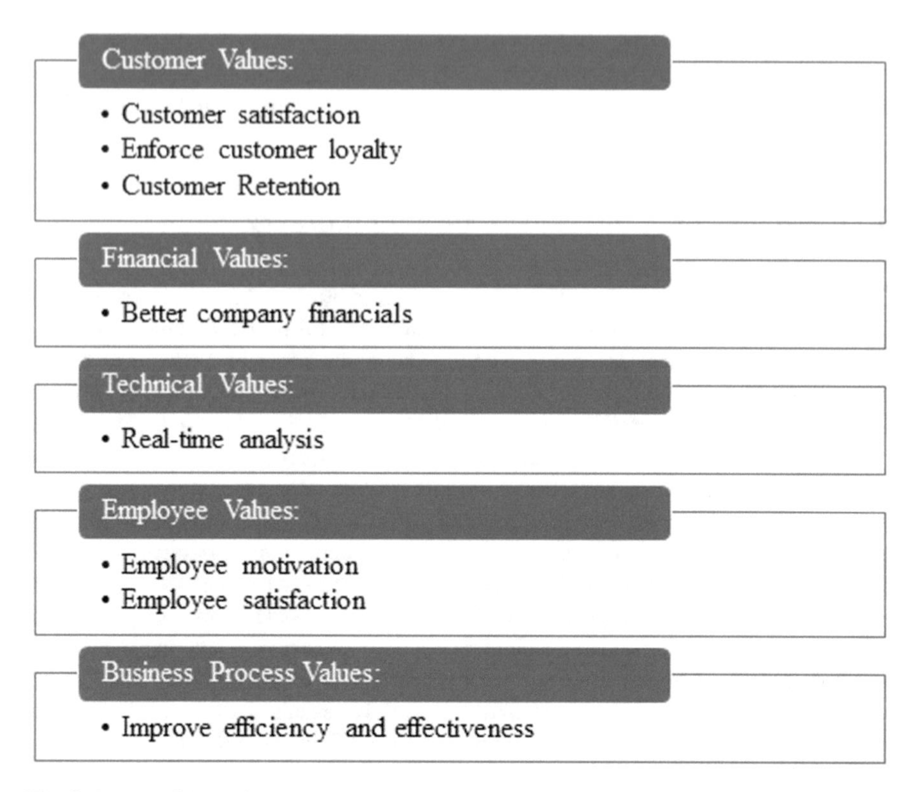

Fig. 3 Impact of BI on CRM

These KPIs need continuous assessments to find out any change in business conditions, increase in opportunities and threats from market competitions.

2. Identifying proper data sources which can give the best intelligence. Few of the sources to derive data are enterprise resource planning applications, shadow systems, content management software and accounting software systems. Social media is another source for unstructured data.
3. Proper testing and cleaning of data to get a better result. All the valuable information hidden in the structured and unstructured data can only be mined properly if combined in the right way. The data sources should be tested properly for quality, and a cleaning and analyzing tool should be selected. The data should be broken down into smaller matrices for data visualization and charting.
4. Keeping proper track of activities on the tools as the number of users grows to comprehend if the information is helpful. Periodical changes should be made in the dashboard, KPI's and the report to keep the system updated with the current trends.

Among all the business sectors using CRM and BI, telecommunication industry is one sector with the maximum number of consumers and generates data in billions which includes customers' call record details. In order to gain competitive advantages over their market competition, the usage of BI is very important in telecommunication industry and also in CRM system used in telecom to understand their customers [15]. Telecommunication is one of the fastest progressing business sectors where customers are continuously switching subscription, so while some are going through termination, new customers are coming in simultaneously. In companies, where it is so easy for customers to terminate their subscription and move on to a different one, customer retention is one of the vital requirements and hence the need for new methods. This industry, as stated earlier, generates a massive amount of data which is continuously increasing in numbers and is more than any other business sector. But all these data would be useless unless properly analyzed that is also necessary for understanding the customers' behavior. This is why telecommunication sector requires proper BI and CRM tools in order to stay ahead in market competition. Some advantages of using BI influenced CRM in telecommunication industry are: customer satisfaction, customer and brand loyalty, customer retention, reduced cost and increased revenue, better data quality, improved customer services and business processes [16].

There are mainly three types of data available in the telecommunication industry. Firstly, customers' call details record (CDR) which is an exhaustive record of all the calls that are being made by the subscribers. CDRs are real-time data used for different purposes, one of them being billing. This record can show the calling behavior of a customer. The features that are recorded under the CDR are average call duration, percentage of unanswered calls, percentage of calls from/to a different area code, percentage of weekday calls, percentage of daytime calls, average number of calls received per day, average number of calls originated per day, and number of unique area codes called over certain time period. Secondly, customer data which includes mainly the demographic information about the customer like name, address, service plan and contract information, credit score, family income and payment history. Since the number of customers in the industry are in millions, keeping record of the customers is vital and hence the necessity to store the records. Lastly, the network data generated by the thousands of interconnected nodes forming the network configuration. These data are generally error or status messages used for the network management purpose [17, 18].

4 Security Challenges Pertinent to CRM and Telecommunication Industry

Many major organizations are victims of security breaches. When it comes to working with huge amount of data, security becomes the major concern of every organization. From the big data perspective, the sensitive data of the organizations should be integrated within the bigger data. To be able to do this, companies need to start implementing self-configurable security policies. These strategies should influence the current existing trust relationship and encourage resource and data sharing inside the organizations [19]. All of these policies should be done while making sure that the big data analytics are properly developed and not restricted because of such strategies. In certain scenarios where if only one of the sites is attacked because of hacking and other security breach, multiple clients would be prone to service disruption. Such threat can be lessened by implementing proper security applications, encrypted file systems, data loss software and security hardware for manual tracking [20].

Even though the data in the CRM application is the most crucial asset of a company, due to its key role in everyday business activities, it is often susceptible to security breaches and disturbances. This is where the concept of information security (IS) and data security comes into play. IS can be defined as the safety of information systems against intruders and unauthorized access to or alteration of information from storage, processing or transit and against rejection of access to authorized users. It also includes the actions required to detect, document and counter these threats. IS has always concentrated on the confidentiality of information, and now with fast growth of data, organizations need to increase their security to safe-keep the privacy of the information and prevent fraudulent activities [21, 22].

The concept of 'privacy' includes human self-esteem, degrees of familiarity, social relationships, and unwanted access by others. Although all these conceptualizations are meaningful, in the scenario of CRM, the concept of information privacy is taken into consideration. Privacy is a significant issue in any business sector including telecommunication industry due to high amount of personal data available with these companies and the possibilities of this data being misused are endless. There are four states of privacy: Solitude, Intimacy, Anonymity and Reserve.

The main functions of privacy are [23]:

1. Personal autonomy which is a wish to avoid being exposed or controlled by others.
2. Emotional release is getting rid of stress from the social life.
3. Self-evaluation.
4. Limited and protected communication.

This scenario gives an idea regarding how security breaches would affect information privacy in businesses and how it should be handled. There are three standard components of security: confidentiality, integrity and availability. The idea of security components is different from the perspective of a business organization and its consumers [24]. When it comes to confidentiality, both the sides demand for prevention of unauthorized access of the data. As for integrity, the organizations should use the data provided by the consumers for business purposes only and not sell those personal data to other organizations without authorization, and also, it is the consumers' responsibility to provide accurate data. For availability, the organization must make sure that the data is available for the customers and businesses, and from the customer's point of view, it needs to be available for modification purposes [25, 26].

Some of the major security threats and vulnerabilities come from the internal sources. Figure 4 shows some of the internal threats caused by both the organization and the customers, and Fig. 5 shows some of the external threats toward the organization and customers' side [27].

Telecommunication industry being one of the largest business sectors dealing with millions of customers and huge amount of data is susceptible to the highest amount of security threats. The subscribers should be able to put their trust in a service provider and the services provided by them. As network operators and service providers have a responsibility toward their subscribers, business partners and public to fulfill therefore, there is a requirement to incorporate security to protect their processes and business interests. The resources that require protection are communication and

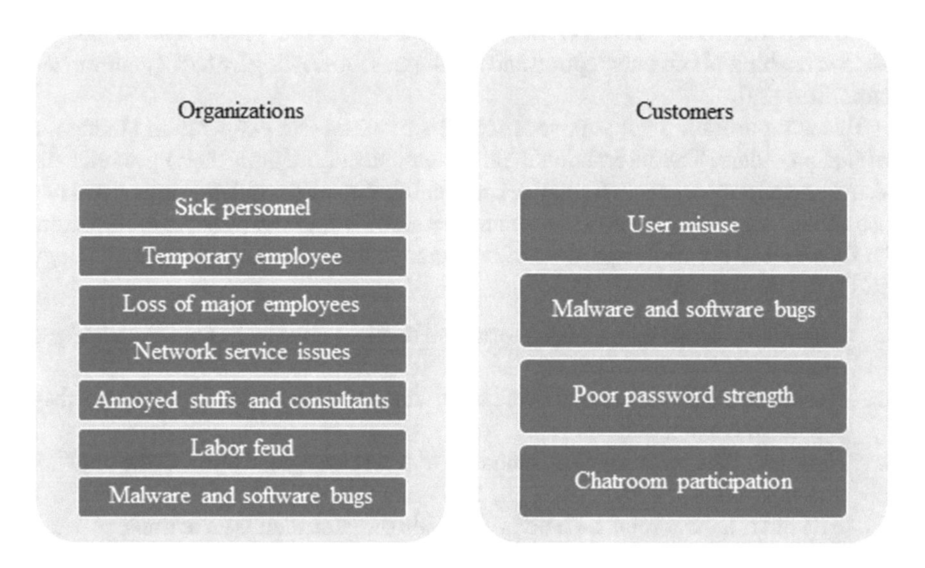

Fig. 4 Internal threats on organizations and customer

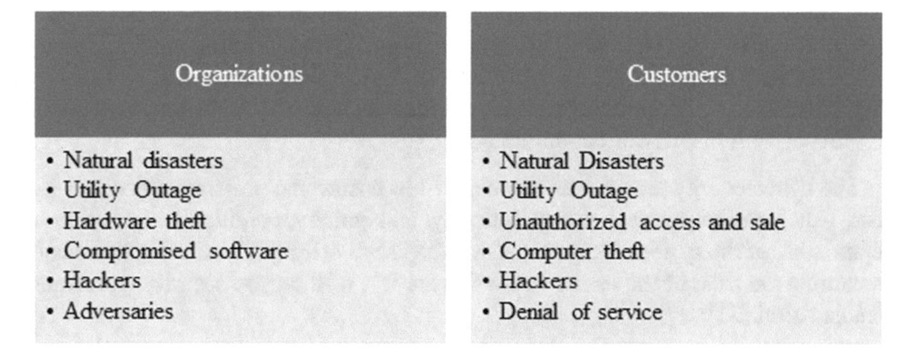

Fig. 5 External threats on organizations and customer

computing services, information and data and software related to the security system [28].

4.1 Preventive Measures Against the Security Issues

The available security services can offer protection to the telecommunication networks against outside malevolent attacks like service denial, message tampering, forgery, hacking and so on. Proper prevention, detection and recovery from outside attacks alongside management of information regarding security are the protection techniques provided by the security services. These techniques must provide protection from both natural disasters and malicious attacks. Also, required steps must be taken to enable smooth interception and continuous monitoring by legally authorized authorities [29].

Telecommunication network security calls for extensive cooperation among the service providers. The International Telecommunication Union (ITU) provides the security requirements and a framework to identify threats toward the networks. They also administer the guidance of the countermeasure plans that can reduce the risk from the threats. The security objectives of information and communications technology (ICT) networks are [30]:

1. Admission to, and usage of networks and services must be limited to authorized users only;
2. Authorized users must be allowed admittance and to operate on resources they are allowed to access;
3. Networks must provide confidentiality to the point agreed upon in the network security policies;
4. All network units must be held answerable only for their own actions;
5. Networks must be secured against unwanted access and illegal operations;
6. Security-related information must be accessible only to authorized users via the network;
7. Strategies must be in place to address how security incidents are to be handled;
8. Procedures must be in place to reinstate usual operation following detection of a security breach;
9. Different security policies and security mechanisms of varied strengths must be at par with the network configuration.

The cybersecurity techniques can be used to ensure the previous objectives are met. It helps with the availability, integrity, authenticity, confidentiality and non-repudiation of the system while respecting the user's privacy. These techniques help in earning the trust of the users. Following are few of the cybersecurity techniques that can used [31]:

- Cryptography provides confidentiality to data.
- Access controls, prevents unauthorized access.

- System and data integrity prevents unauthorized data modification.
- Audit, logging and monitoring evaluate the effectiveness of the security strategies.
- Security management contains risk management, security configuration and controls, incident handling and management of security information.

These are few of the suggested techniques that can be adopted by the industry to ensure the safety of the vital information store with them. New and more efficient techniques keep being developed so it is necessary for the industry to keep themselves updated to ensure highest form of security.

4.2 Proposed Framework

The security framework in Fig. 6 is a schematic representation of incorporating security in CRM and BI for telecommunication industry. In the front-end of the CRM, customers need to be authenticated before access is granted for any kind of information retrieval or modification. Customers, once authenticated, can access their data through different customers' touch points like call centers, email, Web portal and so on. Counter measures should be available within the internal framework of the access control system such that authorized users are not faced with the problem of service denial to their personal data.

The back-end office of CRM stores all the data collected from both the internal and external sources. The internal sources are mostly structured and are behavioral and demographic data of the customers, whereas the external data is mainly unstructured and includes customers' feedback collected through online platforms or call centers. These data need to be encrypted using encryption techniques to maintain customer privacy and data integrity. The encryption techniques used can either be asymmetric or symmetric encryption. In asymmetric encryption, a public and a private key is used for the purpose of encryption, but in symmetric, a single key is used. Rivest-Shamir-Adleman (RSA) is an example of asymmetric encryption algorithm, and advanced encryption standard (AES) is an example symmetric encryption algorithm.

5 Conclusion

In this chapter, we have discussed the importance of security in business sectors that handles an ever-increasing amount of data every day and the impact of security breach that can have a domino effect on CRM. Big data analytics plays an integral part in better decision-making in BI, and this in turn influences the CRM processes to get a better understanding of the customers' behavioral pattern. Telecommunication industry uses these processes to improve their relationship with the customers and retain them. But along with data and information come the security challenges which can be the biggest threat to a network operator. The privacy of the customers should be

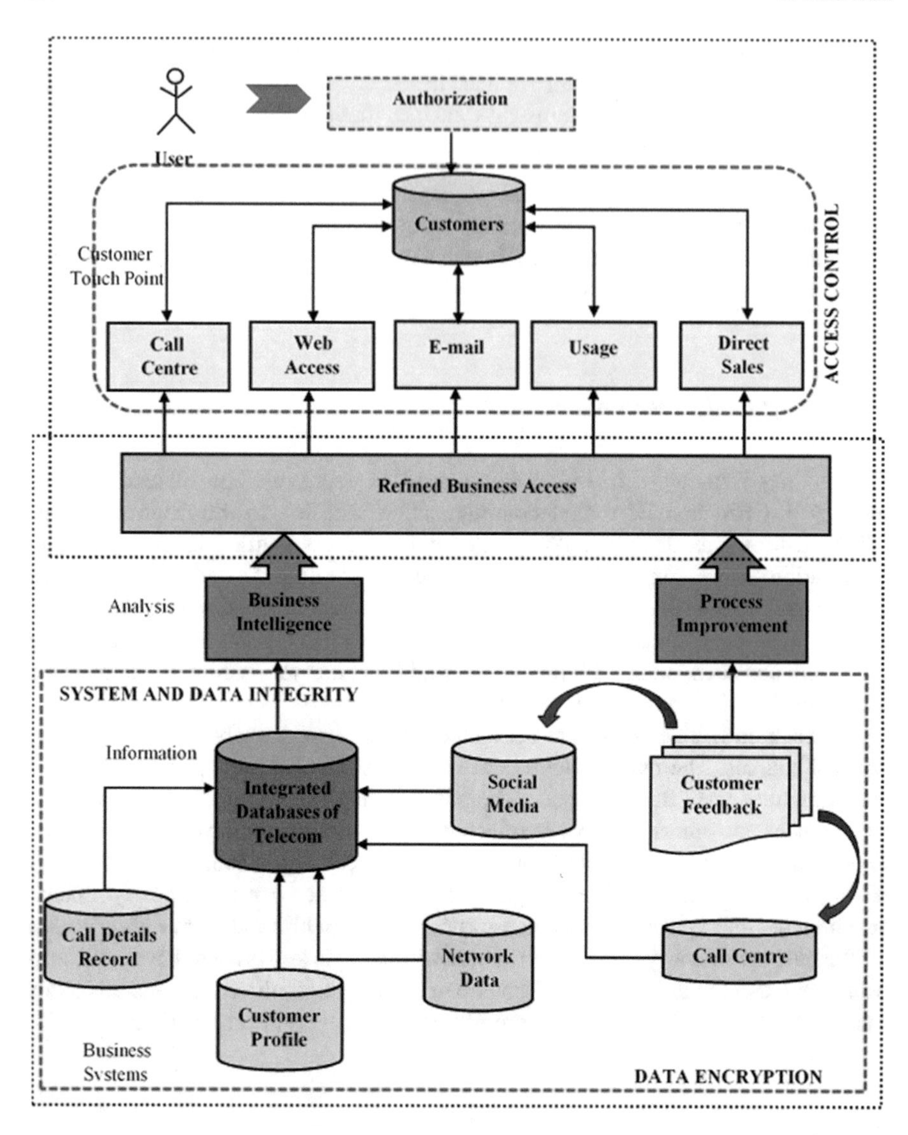

Fig. 6 Proposed framework for security implementation in CRM and BI for telecommunication industry

maintained by protecting the information provided by them in order to avoid customer churn due to loss of trust in the service providers. Different security techniques are implemented to protect the information and the privacy of their customers. In order to address the security challenges, a framework has been proposed that would ensure privacy preservation by safeguarding various information through different data encryption techniques.

References

1. Mikalef P, Pappas IO, Krogstie J, Pavlou PA (2020) Big data and business analytics: A research agenda for realizing business value. Inf Manag 57(1):103237
2. Krishnamoorthi S, Mathew SK (2018) Business analytics and business value: a comparative case study. Inf Manag 55(5):643–666
3. Dong JQ, Yang CH (2020) Business value of big data analytics: a systems-theoretic approach and empirical test. Inf Manag 57(1):103124
4. Fan S, Lau RY, Zhao JL (2015) Demystifying big data analytics for business intelligence through the lens of marketing mix. Big Data Res 2(1):28–32
5. Ram J, Zhang C, Koronios A (2016) The implications of big data analytics on business intelligence: a qualitative study in China. Procedia Comput Sci 87:221–226
6. Jukic N, Jukic B, Meamber L, Nezlek G (2002) Improving e-business customer relationship management systems with multilevel secure data models. In: Proceedings of the 35th annual Hawaii international conference on system sciences. IEEE, pp 2256–2265
7. Baashar Y, Alhussian H, Patel A, Alkawsi G, Alzahrani AI, Alfarraj O, Hayder G (2020) Customer relationship management systems (CRMS) in the healthcare environment: a systematic literature review. Comput Stand Interfaces, p 103442
8. Sun Z, Sun L, Strang K (2018) Big data analytics services for enhancing business intelligence. J Comput Inf Syst 58(2):162–169
9. Dedić N, Stanier C (2016) Towards differentiating business intelligence, big data, data analytics and knowledge discovery. In: International conference on enterprise resource planning systems, Springer, Cham, pp 114–122
10. Balachandran BM, Prasad S (2017) Challenges and benefits of deploying big data analytics in the cloud for business intelligence. Procedia Comput Sci 112:1112–1122
11. Phan DD, Vogel DR (2010) A model of customer relationship management and business intelligence systems for catalogue and online retailers. Inf Manag 47(2):69–77
12. Handzic M, Ozlen K, Durmic N (2014) Improving customer relationship management through business intelligence. J Inf Knowl Manage 13(02):1450015
13. Mishra S, Mishra BK, Tripathy HK, Mishra M, Panda B (2018) Use of social network analysis in telecommunication domain. In: Modern technologies for big data classification and clustering, IGI Global, pp 152–178
14. Panda B, Mishra S, Mishra BK A meta-model implementation with tabu search technique to determine the buying pattern of online customers. Ind J Sci Technol 9:S1
15. Dutta A, Misra C, Barik RK, Mishra S (2021) Enhancing mist assisted cloud computing toward secure and scalable architecture for smart healthcare. In: Hura G, Singh A, Siong Hoe L (eds) Advances in communication and computational technology. lecture notes in electrical engineering, vol 668. Springer, Singapore. https://doi.org/10.1007/978-981-15-5341-7_116
16. Al-Zadjali M, Al-Busaidi KA (2018) Empowering CRM through business intelligence applications: a study in the telecommunications sector. Int J Knowl Manage (IJKM) 14(4):68–87
17. Weiss GM (2005) Data mining in telecommunications. In: Data mining and knowledge discovery handbook. Springer, Boston, MA, pp 1189–1201
18. Mishra S, Mallick PK, Jena L, Chae GS (2020) Optimization of skewed data using sampling-based preprocessing approach. Front Publ Health 8:274. https://doi.org/10.3389/fpubh.2020.00274
19. Mishra S, Mahanty, C, Dash S, Mishra BK (2019) Implementation of BFS-NB hybrid model in intrusion detection system. In: Recent developments in machine learning and data analytics. Springer, Singapore, pp 167–175
20. Saha L, Sahoo L, Routray SK (2017) Survey on CRM analytics in telecommunication industry. Int J Comput Trends Technol (IJCTT) 45(1):21–27
21. Al-Aqrabi H, Liu L, Hill R, Ding Z, Antonopoulos N (2013) Business intelligence security on the clouds: challenges, solutions and future directions. In: 2013 IEEE seventh international symposium on service-oriented system engineering, IEEE, pp 137–144

22. Kim S (2010) Assessment on security risks of customer relationship management systems. Int J Software Eng Knowl Eng 20(01):103–109
23. Mishra S, Tripathy HK, Mallick PK, Bhoi AK, Barsocchi P (2020) EAGA-MLP—an enhanced and adaptive hybrid classification model for diabetes diagnosis. Sensors 20(14):4036
24. Mishra S, Sahoo S, Mishra BK (2019) Addressing security issues and standards in Internet of things. In: Emerging trends and applications in cognitive computing. IGI Global, pp 224–257
25. Rath M, Mishra S (2020) Security approaches in machine learning for satellite communication. In: Machine learning and data mining in aerospace technology, Springer, Cham, pp 189–204
26. Rath M, Mishra S (2019) Advanced-level security in network and real-time applications using machine learning approaches. In: Machine learning and cognitive science applications in cyber security. IGI Global, pp 84–104
27. Romano NC Jr, Fjermestad J (2007) Privacy and security in the age of electronic customer relationship management. Int J Inf Secur Priv (IJISP) 1(1):65–86
28. Martucci LA, Zuccato A, Smeets B, Habib SM, Johansson T, Shahmehri N (2012) Privacy, security and trust in cloud computing: the perspective of the telecommunication industry. In: 2012 9th International conference on ubiquitous intelligence and computing and 9th international conference on autonomic and trusted computing, IEEE, pp 627–632
29. Mishra S, Tripathy, N, Mishra BK, Mahanty C (2019) Analysis of security issues in cloud environment. Secur Des Cloud, Iot, Soc Network 19–41
30. HandbooN, I.T.U.T (2003) Security in telecommunications and information technology–an overview of issues and the deployment of existing ITU-T recommendations for secure telecommunications
31. Mallick PK, Mishra S, Chae GS (2020) Digital media news categorization using Bernoulli document model for web content convergence. Pers Ubiquit Comput. https://doi.org/10.1007/s00779-020-01461-9

Chapter 11
Data Protection and Data Privacy Act for BIG DATA Governance

Kesava Pillai Rajadorai, Vazeerudeen Abdul Hameed, and Selvakumar Samuel

1 Introduction

In this digital age, reaction to the information technology is becoming much more complex. As the technology is reaching a maturity level, new businesses challenge each other to gain quick profits. In this context, the main victim of this war is personal data. Organizations are illegally trading personal data of individuals and organizations without getting consent from the related parties. As the reason, all the countries are starting to worry about the confidentiality and privacy of individuals' data. The existence of "Big Data" makes the situation worst.

Big data is a collection of huge size of gigantic and multifaceted datasets and data volume that include the huge volume of data, data manipulation capabilities, multidimensional analytics, and real-time data. Big data analytics is the process of exploration massive data. The data volume is measured in terms of terabytes or even petabytes. As the nature of its volume size, no traditional data management tools can able to store and process it. At present big data is the buzz topic in this information world, relating it in multiple domains such as in science, arts, business management, security, national or international policy creation, etc. However, surprisingly in many sectors big data never uses personal data at all. For example, using health and psychology data possibly will ascertain new drugs for pandemic disasters without using personal data. Nevertheless, some sectors still use personal data such as social media, data collecting sensors (e.g., RFID tags). Those sectors should ensure that they do not infringe any data protection acts.

K. P. Rajadorai (✉) · V. A. Hameed · S. Samuel
Asia Pacific University of Technology and Innovation, Kuala Lumpur, Malaysia
e-mail: kesava@staffemail.apu.edu.my

V. A. Hameed
e-mail: vazeer@staffemail.apu.edu.my

S. Samuel
e-mail: selvakumar@staffemail.apu.edu.my

P. K. Das et al. (eds.), *Privacy and Security Issues in Big Data*, Services and Business Process Reengineering, https://doi.org/10.1007/978-981-16-1007-3_11

There are three types of big data: structured, unstructured, and semi-structured. Structured big data can be stored, processed, and retrieve in preferred format, for example, patient records, patients' health data, and their drugs usage can be retrieved in an organized manner and unstructured big data refers to the data have no specific format or structure, making it difficult and complex to stored, processes and analyze. For example, data gathered from surveillance camera or interviewee's comments, whereas, semi-structured big data is the cross-bred of structured and unstructured where, these types of data are not been grouped to any particular database and classified but contain some essential data that can be used to segregate individual elements within the dataset.

Anyway, the volume is not so important but what the business entities are doing with the volume of data that matters. Normal data will reveal the basic information of the data provider, whereas big data can be used to be scrutinized for understandings and to provide tips for enhanced strategic decision making. As a result, implementing data protection and data privacy acts on big data becomes essential to safeguard this complex information from corruption, compromise, or loss. The significance of data protection upsurges as the volume of data increases and stored continues to grow at exceptional rates. As the importance of data protection increases, there will be zero tolerance for data downtime that making it impossible to access important information.

Analytics always uses some types of algorithms to recognize patterns in data that can offer visualized perceptions. For example, as shown in Fig. 1, homes can use analytics to permit innovative functionality that can then transform them into dynamic grid assets in the Energy Cloud. Yet, data protection and privacy become the critical obstacle for these types of resolutions. Lately, data hackings and data intrusions have become common and have caused big-name companies have experienced security breaches of consumer data.

The impact of data protection and data privacy on big data can be comparatively complex from legal aspects. Moreover, some aspects of data protection and data privacy principles, terms and regulations will be very difficult to accommodate in

Fig. 1 Big data house

to the main characteristics of big data. On top of that, the data elicitation procedure for big data at all time is collection from multiple sources and will be used and shared by different stakeholders for different purposes. As the result, the big data characteristics will be very difficult to follow certain legal procedures associated to data protection and privacy. However, it is imperative to investigate how these legal procedures can be deployed in big data analytics practice.

In the effort of adapting data protection and data privacy acts into big data analytics, European Union Legal framework that is General Data Protection Regulation (GDPR) will be used in this study. However, only the main principles and concepts which used by GDPR which associated many stakeholders dealing with big data analytics will be examined and to reconcile with complex technologies.

This chapter is emphasis to provide an overview to infuse data protection and data privacy act into big data analytics. In the first part, the introduction of data protection and data privacy acts and explanations in general were discussed. The second part which is the main idea of this chapter is the analysis of how data protection and privacy acts can be used in big data. Some issues and recommendation will be discussed using the core principles and concepts of GDPR that many stakeholders active in the domain of big data analytics.

2 Data Protection and Data Privacy Act

2.1 Introduction to Data Protection and Data Privacy Act

In the current digital age, controversy adjoining how public protect their privacy of individual from others has become the international concerns. Moreover, the development of the information and communication technology (ICT) provides path to the vast use of computer and information technology. Secondly, with the emergence of data storage and repository technology, data was greatly stored in the hard disk for personal computer (PC) and in cloud storage for portability. Thirdly, significant invention of Internet services, social networks, Internet of things (IoT), and cloud systems uses personal data [2]. On top of that, the creation of fake identities on social Web sites directly endangers the individual and corporates, due to the fact that personal data becomes easily accessed by anyone and become the critical concern among the people. Furthermore, privacy is a basic right preserved in many constitutions around the world. Privacy is complex to define as it has multiple agenda to people but need to understand the fundamental aspect of it, as it affects public's individual confidentiality [1]. At the moment, current jurisdictive model prevails complex in terms of technicality and inconsistent cybersecurity and data privacy law. This is where the concern for the protection of personal data begins.

2.2 Personal Data

Every individual and public has the right to keep their personal information as private and confidential. As well as most countries have their own jurisdictive model to ensure that the personal data of an individual is protected to maintain the consistency of security and data privacy. European Commission and Commission Nationale de l'Informatique et des Libertés of Personal and GDPR is define as "Personal data is any information that relates to an identified or identifiable living individual. Different pieces of information such as identification number (e.g., social security number) or one or more factors specific to his physical, physiological, mental, economic, cultural or social identity (e.g., name and first name, date of birth, biometrics data, fingerprints, DNA…, which collected together can lead to the identification of a particular person, also constitute personal data" [3, 4].

Figure 2 shows the types of personal data. Generally, personal data can be categorized as sensitive personal data and general personal data, where data that harm a person's daily life or endanger or life threatening can be considered as sensitive personal data, for example, personal bank account information, health data, national identification number, etc. On the other hand, personal data which a person uses day

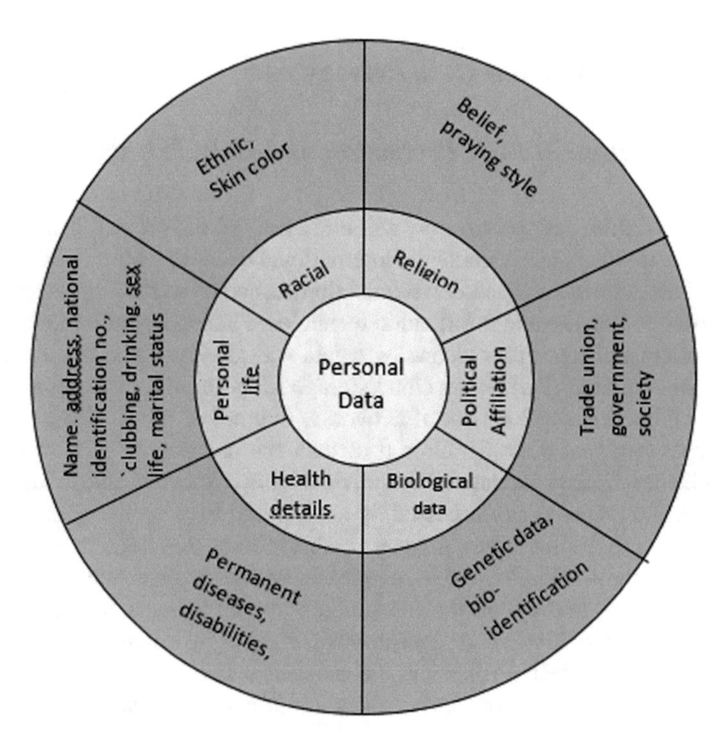

Fig. 2 Type of personal data

to day without fear, such as name and address, car, qualifications, and other information which considered as non-sensitive will fall under general personal data. Apart from that, the data that used for legal purpose such as legal person's name, e-mail, financial status, and marital status is not considered as personal data of an individual.

3 Data Protection

Every human in this world has the legal right to protect his personal data without revealing to the public or to anyone. In this context, every country has certain legal procedure of its own to maintain and to protect the privacy and confidentiality of their citizens' personal data. Data protection is generally explained as the commandment designed to protect your personal data. In edict to endow us to manage our personal data and to protect us from manipulations and exploitation, it is crucial that data protection laws confine and shape the actions of companies and governments. For example, in this digital era, our daily life and ICT become inseparable, where data and information are the most required entity to process our daily activity. Therefore, revealing our personal data becomes unavoidable. Using these opportunity, some unethical individuals and organizations are abusing this data to gain short-term and long-term profits. For instance, some financial institutions have millions of consumers' personal data, illegally selling the data to other servicing organization and gaining millions of dollars. On the other side, many scammers pray the innocent public to cheat them and stealing their life saving [5]. Table 1 shows the statistical analysis of victims and the amount of money they lost in scam.

3.1 Purpose of Data Protection

Every time you use a service, buy a product online, register for e-mail, go to your doctor, pay your taxes, or enter into any contract or service request, you have to hand over some of your personal data. Once the transaction is over, the personal data will be exposed. Do we ever think, what will happen to the data after they collect and processed? The companies and organizations that you deal with will save the data and information for their future use such as when advertising their product and services or even they will sell to the unknown agencies. With your knowledge or not, the data and information about you are being become widespread. The worst case is when your information were used by some agencies and organization that you may not even done any transaction, will be using them for their own benefits.

To avoid all these unfavorable situations of citizens and consumers and to increase the confidence in both public and private sectors, strong data protection practices and acts were introduced. With the effective legislation, the government can now be able to control and help to reduce the countries' surveillance and data misuse.

Table 1 Top 10 scam types in 2019. (*Source* SPF) [5]

	Cases reported		Amount cheated		
	2019	Change from 2018	2019	Change from 2018	Largest sum cheated
E-commerce scams	2809	+ 648	$2.3 m	+ $400 k	$180 k
Loan scams	1772	+ 805	$6.8 m	+ $4.9 m	$195 k
Credit for-sex scams	1065	+ 532	$2.8 m	+ $1.3 m	$80 k
Social media impersonation scams	810	+ 703	$3.1 m	+ $2 m	$330 k
Internet love scams	649	−19	$34.6 m	+ $6.7 m	$4.6 m
Investment scams	508	+ 161	$36.9 m	−$2.5 m	$5.3 m
China official impersonation scams	455	+ 154	$21.1 m	+ $8.4 m	$1.3 m
Business e-mail impersonation scams	385	+ 10	$45.4 m	−$11.4 m	$4.1 m
Lucky draw scams	311	+ 152	$1.2 m	−$600 k	$314 k
Tech support scams	249	+ 226	$13.9 m	+ $13.8 m	$1.6 m
Total	9013	+ 3372	$168 m	+ $23 m	

3.2 *Principles of Data Protection*

To protect the personal data of individuals from vulnerability and misinterpret GDPR as the Data Protection Directive (DPD) has designed seven data protection principles to safeguard the personal data. The principles are as follows (Table 2):

- Lawfulness, fairness, and transparency

Data collector who collects the personal data of individual should be ethically gathered, processed in brief, transparent, and clear, which can be easily available, and in simple language. This principle is very useful when someone fraudulently sell or transferring of personal data.

- Purpose limitation

Data controller handling a personal data should be limited to the purpose when it being acquired. Further processing of the same personal data in the future for

Table 2 Seven principles of data protection [8]

Lawfulness, fairness and transparency	Personal data shall be processed lawfully, fairy and in a transparent manner in relation to the data subject
Purpose limitation	Personal data shall be collected for specified, explicit and legitimate purposes and not further processed in a manner that is incompatible with those purposes
Data minimisation	Personal data shall be adequate, relevant and limited to what is necessary in relation to the purposes for which they are processed
Accuracy	Personal data shall br accurate and, where necessary, kept up to date
Storage limitation	Personal data shall be kept in a form which permits identification of data subjects for no longer than is necessary for the purposes for which the personel data are processe
Integrity and confidentiallity	Personal data shall be processed in a manner that ensures appropriate security of the personal data, including protection against unaythoriised or unlawful processing and against accidental loss, destruction or damage, using appropriate technical or organisational measures
Accountability	The controller shall be responsible for, and be able to demonstrate compliance with the GDPR

"other purposes" require authorized permission again from the owner. However the exception is, the "other purpose" is similar to the original purpose.

- Data minimization

Under the GDPR principle, the data controller should make sure that personal data is used only when necessary and it should be very specific. Personal data should be explicitly explained. Data controller should ensure the personal data used is for the specific purpose and should not deviate from the purpose.

- Accuracy

Personal data collected must be accurate, current, and correct format according to the DPD. Data controller should be responsible for the maintaining and managing personal data. Any fuzzy, incomplete, inaccurate, misleading, and outdated data must be removed or corrected.

- Storage limitation

As storages are limited data, controller must be responsible to manage the usability of the personal data. When a personal data is no longer in use for the purpose for which is collected must be deleted or if there is a reason for retaining it then it should be archived. There must be a regular review using standard operating procedure (SOP) for reviewing and managing the data repository.

- Integrity and confidentiality (security)

Data must be protected against vulnerability or illegal access using appropriate approaches. Data controllers must review the risk of implementing certain appropriate security measures for the data and decisively review the procedures are up-to-date and working efficiently and effectively. To safeguard the integrity and confidentiality of the personal data, the government of each country should implement a strict breach reporting system.

- Accountability

It is about compliance burden, means who is going to be responsible for any disaster. It has very complex processes that need to put in place to guarantee the data is secure. Some of the processes are as follows:

- Creating and evaluating existing data privacy processes
- Appointing data protection officer
- Creating personal data repository
- Deploying appropriate privacy notices
- Getting consents from owners
- Implementing proper approaches and measures to ensure compliances
- Creating a specific metrics for privacy impact assessment (PIA)
- Implementing a breach reporting system.

3.3 Available Data Protection and Data Privacy Acts

The following are some major data protection and data privacy acts that are implemented in USA and UK (Table 3) (Fig. 3).

3.4 Current Data Protection Practices

4 Data Privacy

Privacy is defined as a situation where any individual is not been observed or intrude by anyone else. Therefore, the privacy needs to be recognized and protected. In this context, the international human right Article 12 of the Universal Declaration of Human Rights (UDHR) declare that.

Table 3 Data protection and data privacy act

Year	Data privacy and acts	Description
1970	U.S. Fair Credit Reporting Act	To help the accuracy, fairness, and privacy of consumer data and envisioned to safeguard consumers from the deliberate and/or inattentive inclusion of wrong information in their credit report
1974	U.S. Privacy Act	Creates a standard of reasonable information processes that administrates the acquiring, preservation, use, and distributing of information about individuals that is maintained in systems of records by federal agencies
1986	U.S. Computer Fraud and Abuse Act	This act bans retrieving or broadcasting personal or organization private information from computer without authorization
1986	U.S. Electronic Communications Privacy Act (ECPA)	Addresses seizure of dialogues using fixed telephone lines, however it did not apply any ICT devices. Nevertheless, other acts, such as USA Patriot Act, are covering the evolution of new technology
1987	U.S. Computer Security Act	Addresses the establish standard and guidelines under National Institute of Standards and Technology (NIST), creates customized security plan (System Security Authorization Agreement (SSAA)) processes for processing sensitive information. It also involves user information security training, assessing vulnerability of federal computer system, rendering technical assistance joining with National Security Agency (NSA) and creating training guidelines for federal personnel
1988	U.S. Video Privacy Protection Act	These acts protect the users or consumers from wrongful expose of an personal identifiable information restricting from their rental or purchase of audiovisual material, including videotapes, DVDs, and video games
1990	United Kingdom Computer Misuse Act	This acts targeted to protect consumers from computer misuse offences such as unauthorized access, to computer materials, intent to harm operation of computer, etc

(continued)

Table 3 (continued)

Year	Data privacy and acts	Description
1995	Data Protection for the European Union (EU)	It is a council directive—on safeguarding the individuals with regard to personal data processing and the distribution of these data and it has obliged to the independent movement of personal data and regulated the process collection and processing in states of the European Union
1996	HIPAA—Health Insurance Portability and Accountability Act	It is a federal law of Privacy Rule to protect individuals' health information while allowing the flow of health information desired to provide high quality healthcare and wellbeing. It allows to use the important information, while protecting the privacy of people. This act is devised for the diversified and comprehensive healthcare market to protect the various uses and disclosures that need to be addressed
1998	U.S. Digital Millennium Copyright Act (DMCA)	This acts is to protect the inventors/innovators from those forbids making and distributing of technology, strategies, devices, or services for the purpose to evade the copyrighted work which also known as digital rights management (DRM)
1999	U.S. Uniform Computer Information Transactions Act (UCITA)	The purpose of this act is to control deals related to computer product such as software, online databases, and software licensing. Furthermore, It is envisioned to standardize the rubrics to all information communication and technology (ICT) transactions
2000	COPPA—Children's Online Privacy Protection Act	This act applies to the online collection of personal information of children under 13 years of age. It is about the procedure of when and how to seek verifiable consent from a parent or guardian and the operators' responsibilities to protect children's privacy and safety online personal information
2002	FISMA—Federal Information Security Management Act	This act "requires each federal agency to develop, document, and implement an agency-wide program to provide information security for the information and systems that support the operations and assets of the agency, including those provided or managed by another agency, contractor, or other sources" [18]

(continued)

Table 3 (continued)

Year	Data privacy and acts	Description
2013	ISO 27001	It "specifies the requirements for establishing, implementing, maintaining and continually improving an information security management system within the context of the organization. It also includes requirements for the assessment and treatment of information security risks tailored to the needs of the organization" [19]
2018	GDPR—General Data Privacy Regulation	This acts principally aims to allow individuals to control over their personal data and to simplify the process for international business by standardizing the laws among European Union
2020	CCPA—California Consumer Privacy Act	This act provides consumers the right to reject, access or delete, and sale their personal information

Privacy is an internationally recognized human right. Article 12 of the Universal Declaration of Human Rights (UDHR) proclaims that

> No one shall be subjected to arbitrary interference with his privacy, family, home or correspondence .... Everyone has the right to the protection of the law against such interference or attacks. [6]

4.1 Privacy Policy

Every organizations and agencies who are dealing with consumers need to explain and declare how their personal data is being controlled. Therefore, the privacy policy must be always accessible to the consumers and must be written in layman terms. In other word, the privacy policy should be written in known language and should not have any technical jargons or terminology which have more than one meaning. Almost all countries in the world have their own privacy policies intact and failure to comply can be resulted in heavy penalty [7]. As shown in Fig. 4, in the financial year 2019, USA has collected USD3.92 million through failure to comply data privacy policy.

4.2 Data Protection Through Data Security

Data protection and data privacy are all about protecting personal or organizational data from stealing, hacking, illegal selling, and distributing unauthorized access, whereas protecting information or data from data breaches, cyberattacks, and accidental or incidental data loss by using technologies, techniques, and strategies are the core function of data security [9].

4.3 Data Security Technologies for Data Protection

1. Encryption
   Data and information can be protected by data security technologies such as backups, data masking, and data erasure. The pioneer data security technology is encryption, where data and information are encrypted or manipulate to an unreadable format to prevent from unauthorized users and hackers [10].
2. Authentication
   Another most common data security method is authentication. Data controller will provide a password, personal identification number (PIN), biometric, or other forms of data to verify identity before providing permission to access the data [11].
3. Deceptive Network Technology (DNT) [12]
   When hackers manage to gain entry on a consumers' database, they will start to collect the personal data on the database. This new technology "Deceptive Network Technology" will confuse their search while alerting the individuals or organizations that they have unwanted guest. This technology is similar as booby-trapped or creates ghost network devices such as fake system and servers.
   "Illusive networks," is also another DNT where the technology focuses on splash a consumers' network with a widespread of virtual data.
   "Shadow Networks," technology will create fake databases, servers and system which are similar to the original software to confuse and trap the hackers.

4.4 Existing Data Protection and Data Privacy Acts Policies Acts for Big Data

Big data which includes multiple aspects of data combined to provide a strategic decision for the business is not considered as personal data. According to Article 4(1), GDPR, big data that include personal data, should legislate data protection law, consequently big data analytics which consists of some parts of personal information must also consider data protection legislation in specific to GDPR [13].

Furthermore, the data protection agencies accept the above view, considered big data is also fall in the parameter of data protection law, therefore must observe data protection regulation and GDPR. Moreover, the GDPR was drafted with the idea that the technologies and tools can able to protect the multilayered and multithreaded data [14].

Data privacy is about how data collectors are handling the privacy of a persons' or organizations' personal data such as permission, notification, and governing commitments. In summary, data privacy is all about sharing private data with others, acquiring process and how data collectors observing data protection act.

Moreover, data privacy rules and regulation differ for each domain such as health care and finance [15]. In healthcare industry, data privacy act for European Union is General Data Protection Regulation (GDPR), Health industry is Health insurance portability and Accountability Act (HIPAA), whereas for financial institution Gramm-Leach-Bliley Act (GLBA). However, all these privacy acts are ensuring the right of the business and consumers. Following paragraph explains how to ensure data privacy for business and consumers.

For Businesses: [16]

1. Provide awareness by assimilating in-house or outsource training programs on data privacy acts. It should be a part of yearly agenda, especially to new staff.
2. Use free and trustable security tools such as encryption, passwords and virtual private network (VPN). It can reduce attacks and vulnerability furthermore ease of use and install.
3. Must not overconfident and complacency of our network security and underestimate hackers capability and interest on the organization will weaken the data protection
4. Continuous observation of the network for internal attacks from malicious insiders, vulnerable and apprehensive activities will reduce or eliminate outbreaks in early stage.
5. Introduce zero trust model for data privacy where no stakeholders, software or hardware should have default access to an organization's network. This approach makes the verification process mandatory—"Trust but verify" is unconditionally essential for any organizations.

For Consumers: [17]

1. Use multifactor authentication. Use free and trustable security tools such as encryption, passwords, and virtual private network (VPN). It can reduce attacks and vulnerability furthermore ease of use and install and increase security, preferably non-short message services (SMS)-based multifactor authentication (MFA).
2. Having multiple copy of data regular backup will help to restore the data during disaster (e.g., from malicious insiders, vulnerable and apprehensive activities will reduce).
3. Must have detail knowledge about latest technologies and tools such as IoT devices (e.g., smart home devices—personal assistants, such as Amazon Alexa or Google Home) as it can collect a enormous of data without stakeholder knowledge. IoT has been one of the biggest cyber security threat to the world [18].
4. Always be alert of abnormal transactions, request or attracting offers, block these types of dealings.

5 Current Data Security Challenges in Big Data Protection

The technological growth in information technology, devices, and sensors are being used to generate, communicate, and share data through Internet [19]. The organizations are using these data to make effective and efficient business decisions. As the results, the number of data collected enormously increases; however, individuals and organizations have worry over dependability of these data as well as data privacy and security. Commonly, data violation happens because of simple monitoring processes or data protection rules and regulation were not properly deployed. The following paragraph describes the challenges in data privacy and data security in big data. Data security challenges in big data primarily concern of insecure infrastructure, poor data management, and ethics and compliances [20].

5.1 *Challenges in Using Internet of Things (IoT)*

Digital devices and IT infrastructure are mostly used to capture data, which at the moment were designed with no intention of security.

As shown in figure, devices such as mobile phone, personal computers and smart watches and smart home such as intelligent appliances (Wi-Fi, television, security system, etc.) show tremendous evolution conceivable for integrated devices. As shown in Figure-X, data sharing among things become common and without knowing that home owners are sharing their personal information to unknown criminals. This incredible growth of information technology within home with no proper design for data privacy and security eases the cybercriminals to hack and steal data without fear [21].

The existence of new devices such smart watches, cameras and smart televisions, the consumer who have little knowledge about cybercrimes become their prey. The main reason is this IoT developers basically compromise the things with simple authentication or other simple verification.

Another challenge of data privacy and security is the usage of routers which is vulnerable to hacking. The available routers used at home and in organizations are very simple to hacked using the technique called DNS hijacking without the knowledge, where the criminals redirect the personal data to spy or phishing Web site [22].

5.2 *Challenges in Using Hadoop Technology*

Hadoop technology was developed using open-source software for distributed processing of large datasets across different platforms using programming languages.

As it is distributed processing capabilities, it offers supercomputing power and enormous storing capacity. Hadoop technology not only depends on hardware to maintain reliability, the available libraries in this technology are designed in such a way to sense and manage failures at application layer itself [23]. Therefore, it can manage and provide high available services at any one of a cluster of computers.

When big data given an upsurge, the usage of automated powerful tools and its' capabilities is also started to utilizes by organization to acquire, manage, and analyze massive volume of data for their decision-making processes. However, it brings data security threats for organizations as they keep many sensitive data such as financial data such as bank account number or blue print of their businesses and personally identifiable information. It poses dangerous if the hackers manage to hack this data. Therefore, it is very important for Hadoop technology to have high security capabilities for their storage. Following are some Hadoop technology data security issues;

As Hadoop technology started to use by big data as a service (BDaaS), it started to use cloud's ecosystem of tools and applications (Figs. 3, 4, 5, and 6).

As such the data is freely available and becomes the main concern for data security. Many organizations started to use Hadoop technology for data analytics purposes. This become another concern to data security challenges as earlier the data was only for one organization however, now the data becomes accessible to any users across the organization. Some of the challenges are [24] :

1. User authentication
2. Data sharing
3. Access authorization
4. Historical data backup and deletion
5. Data protection on transactions

5.3 Challenges in Using Cloud Computing

Cloud computing is allowing the stakeholders to store any type of data in cloud environment at remote servers by cloud service providers and permitting them to access the data from anywhere as long as there are Internet facilities. Therefore, the data stored in the remote servers for data processing must be managed with extreme care. Using cloud service delivery model such as software as a service (SaaS), platform as a service (PaaS), and infrastructure as a service (IaaS) provides the access to their data using cloud computing. SaaS allows the consumer to run their storing, creating editing, viewing, and copying their personal data on cloud infrastructure transaction using Web browsers, and PaaS helps the consumer to rent hardware, operating system, storage and network through Internet, whereas IaaS ensures consumers by managing the control process, storage, and basic computing resources.

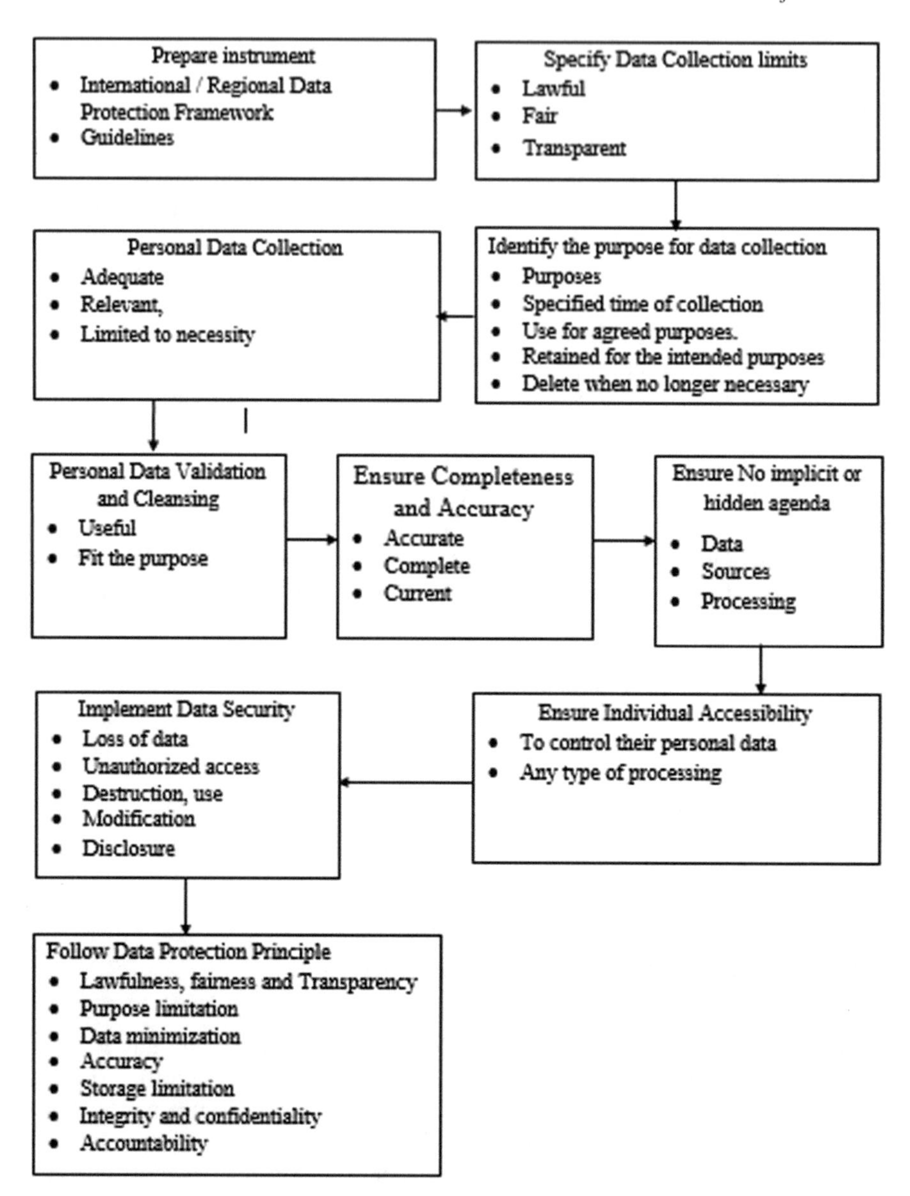

Fig. 3 Data protection practices

Cloud computing can be accessed via a set of cloud computing service models such as software as a service (SaaS), platform as a service (PaaS), and infrastructure as a service (IaaS). In SaaS, the services are provided by the service providers and customers make use of these services to run applications on a cloud infrastructure. These applications can be accessed through Web browsers. PaaS is a way to rent

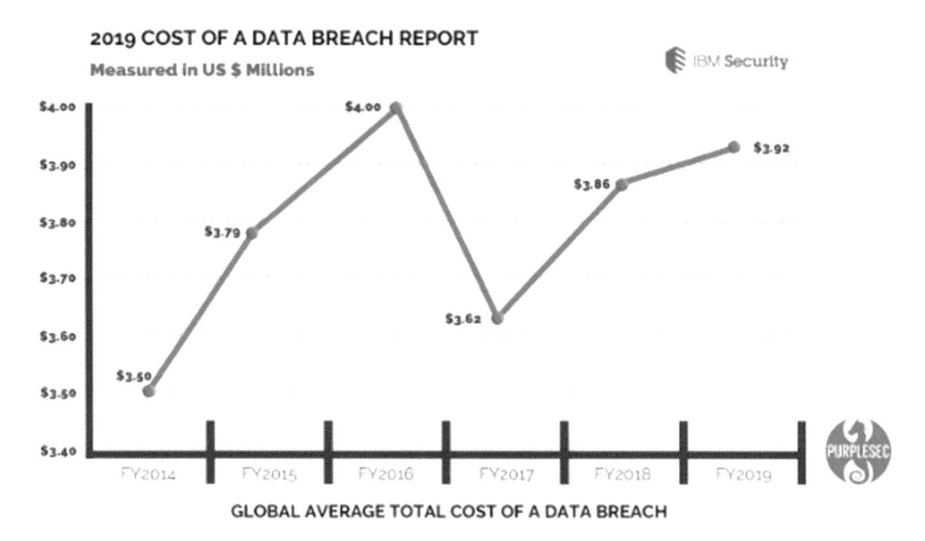

Fig. 4 Cost of a Data Breach Report [7]

hardware, operating systems, storage and network capacity over the Internet. The service delivery model allows the customer to rent virtualized servers and associated services for running existing applications or developing and testing new ones. In IaaS, the consumer is provided with power to control process, manage storage, network, and other fundamental computing resources which are helpful to manage arbitrary software [25].

As cloud computing consumer, we may be thinking that the data is safe. However, this become the biggest challenge for the cloud service providers, where their employees can able to access to the cloud server without permission to misuse the stored sensitive information related to personal data (sensitive or non-sensitive data) [26]. The following are some of the challenges faced by cloud service provider related data privacy and security.

1. Daily data has been captured and will automatically store in cloud storage. Cloud as the virtual storage area, consumers may not realize where the data is being stored and who are the people following and accessing them. Moreover, consumers may not able to identify or control these illegal activities.
2. Cloud data will always move around the clouds to provide easy access to users. Therefore, the users will not able to identify the location of this data, hence data loss will be common [27]. Consumers cannot able to differentiate between authorized and unauthorized users accessing the data. As the data is volatile and move around the virtual machine, it is impossible to have the power to block any access which vulnerable or violating the policies.
3. Every day the number of cloud computing users is increasing exponentially, though the number of data storage is also tremendously increasing. Therefore, managing these enormous number of sensitive (business and individual)

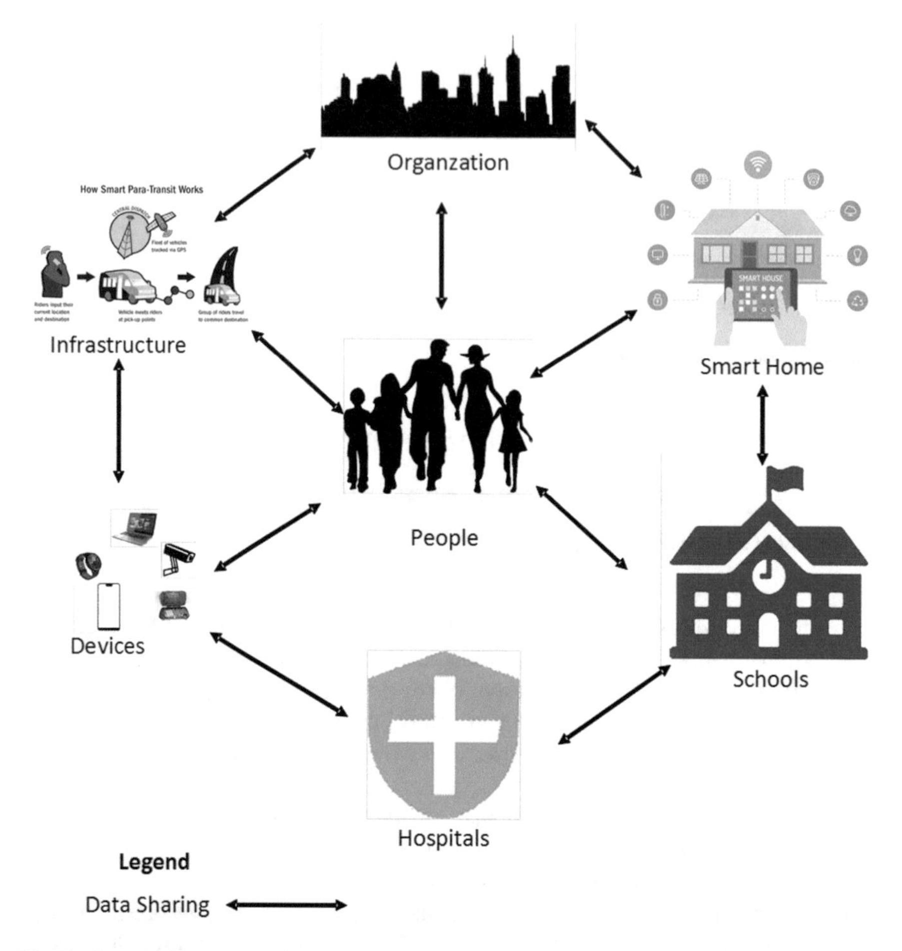

Fig. 5 Data sharing using IoT

data needs many types of resources such as security, storage space, software services, and security administrators. By knowingly or unknowingly any one of the resources went against the procedure or policy the entire cloud will be at risk.

4. Moreover, when many stakeholders share the same cloud for storing, transacting, and processing, there will be high chances for hackers to misappropriate the data. Therefore, data protection will be the biggest challenges for the data administrators. Furthermore, data confidentiality (e.g., illegal access and malware), data integrity (e.g., weak passwords and encryption), and data availability (e.g., data format and data corruption) too facing the risk due to malicious user such as cross-site scripting and access control mechanism [28].
5. Adding to that the usage of cloud server with no standard operating procedures (SOP) has poor data management to save and analysis which correspondingly

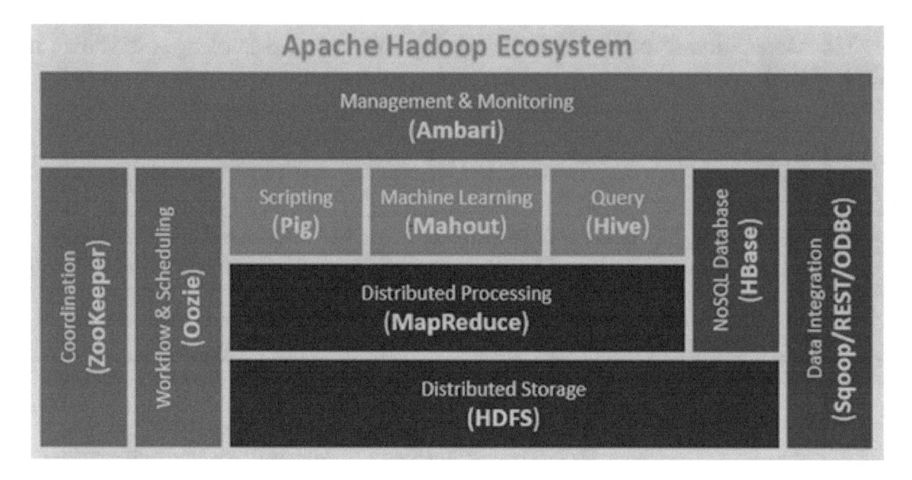

Fig. 6 Apache Hadoop Ecosystem [Image Credit: mssqpltips.com]

increases the challenge to data security, finally societal concern and awareness of data privacy

There are many more data security challenges exist in cloud computing. Some other common challenges depicted in Figs. 3, 4, 5, 6 and 7 [29, 30].

Fig. 7 Other cloud computing data security challenges

The explanation of other cloud computing data security challenges is shown in Table 4 [31].

Table 4 Explanation of other cloud computing data security challenges

Data security challenges	Description
Web browser	There are many Web browsers available. Each and every browsers have their own weaknesses which yet to be solved. The cloud services too use these browsers for their activities where the vulnerability will inject into the cloud servers which will harm the data security and privacy
Access management	It is all about providing access to genuine users according privileges security policy. Therefore, it aims at accomplishing the authentication among various type of clouds to establish an association, however, struggles interoperability issues. Sharing various information among diverse articles which have dissimilar data sensitivity needs vigorous segregation and access control procedure
Malware attack	Injecting malware application into cloud services such as through SaaS or PaaS because certain illegal activity reasons, the hackers or attackers may block, change the original functionalities, and hide the original data or snooping
Compliances	At the moment there are no proper SOP for managing cloud services. Therefore, it has no compliance and prevents management from any serious attacks which can endanger data privacy and security
Software interface	In order to give end user cloud computing services, the cloud service provider designs the interface or API as simple as possible. As the result, security is being compromised
Wicked use	Cloud computing has a huge database and complex computing power, permitting wicked users to attack the infrastructure by spreading malware or malicious processes
Malicious user	It is very difficult to differentiate between legal user and illegal users in cloud computing. Legal user does not mean only staff, but the service providers and also any stakeholders that providing services in that cloud. There is no proper transparency in access processes and SOP to access to cloud assets makes the identification difficult to manage. Therefore, it is very difficult to identify the illegal users from accessing the cloud services
Risk management	Risk is everywhere, as the cloud services yet to have a SOP for managing the cloud services become the ultimate risk, and it is easy to gain the permission to access cloud services without considering security processes or available technology. Therefore, cloud service providers must consider how to acquire, store and use data, accessing the data

6 Data Security Technologies for Data Protection and Data Privacy of Big Data

As mobile technology and businesses cultivate borderless, the personal and business data collection enormously increases, therefore it is important to safeguard the privacy and confidentiality of the data collected from external threats through network. Data protection processes are very complex as information technologies are keep evolving very fast (e.g., cloud application, private/public cloud, data centers, and portable computing), when data users interact across organization. It is imperative to keep up with the latest technologies and rules to protect the privacy of the data from vulnerable attack. Following are some data security technologies that will help data protection and data privacy [32].

1. Data organization
   Organizing data is fundamental principle for data security. Furthermore, in big data management the data is collected various people or organizations. Therefore, it must be organized according to the priority so that it will be easy to store, analyze, and visualized. On top of that, it will be easy to authenticate the originality access. Currently, there are many tools that support data organization.

2. Data authorization procedure
   Allowing precise number of data authorization procedure to users is vital for data protection and data privacy. Allowing only primary users of the organization to manage the sensitive data will ensure to eliminate or reduces data breaches. Providing the right amount of data access to individual users is crucial for data protection.

3. Data encryption and data masking
   Must provide data encryption procedures to data users will protect the sensitive data from unauthorized entries. Moreover, masking the sensitive data too protect from malicious sources. Masking specific areas of data can protect it from disclosure to external malicious sources.

4. Data archive or erase
   Unwanted or old sensitive data must be backed up or erased from cloud databases. This will enable the cloud data administrator to manage the sensitive data efficiently and effectively. By doing this procedure, leakage or data theft can be avoided. Data management can be handy and make accessing easy, safe and ensure availability.

5. Multilayered authentication
   Multilayered authentications are very beneficial for stopping attackers from getting access to sensitive data. Though the hackers manage to access to the first layer, they will struggle to pass through the next layer and following layers. The hackers will be tired of accessing to multilayer authentication, finally they let go the malicious attack.

6. Randomization and volatilizing
   Randomizing or volatilizing the authentication value for sensitive data makes the attackers difficult or confuse to predict the correct value. Randomizing or volatilizing has no fixed precise association between sensitive data and randomized data.

References

1. A guide for policy engagement on data protection, The Keys to Data Protection, August 2018
2. Skendžić A, Kovačić B, Tijan E (2020) General data protection regulation—protection of personal data in an organisation. Polytechnic "Nikola Tesla", Gospić, Croatia, May 2018=-uploaded by Edvard Tijan on 06 February 2019. https://www.researchgate.net/publication/326708317. Access on 23rd May, 2020
3. European Commission. https://ec.europa.eu/info/law/law-topic/data-protection/reform/what-personal-data_en
4. Commission Nationale de l'Informatique et des Libertés. https://www.cnil.fr/en/personal-data-definition
5. Mahmud AH (2020) https://www.channelnewsasia.com/news/singapore/singapore-scam-cases-on-the-rise-crime-rate-12395936. Singapore, "Why scam cases continue to rise and what is being done about them", 05 Feb 2020 03:30 PM (Updated: 05 Feb 2020 06:12PM)
6. GA Res. 217 (III) A, UDHR, art. 12 (Dec. 10, 1948)
7. The Ultimate List of Cyber Security Statistics For 2019. https://purplesec.us/resources/cyber-security-statistics/
8. The data protection principles under the General Data Protection Regulation. https://globaldatahub.taylorwessing.com/article/the-data-protection-principles-under-the-general-data-protection-regulation
9. Follis E (2019) Technology evangelist and consultant—"Data Privacy vs. Data Security: What Is the Real Difference? https://blog.netwrix.com/2019/06/25/data-privacy-vs-data-security-what-is-the-real-difference/. Published: June 25, 2019
10. Mishra S, Mallick PK, Tripathy HK, Bhoi AK, González-Briones A (2020) Performance evaluation of a proposed machine learning model for chronic disease datasets using an integrated attribute evaluator and an improved decision tree classifier. Appl Sci 10(22):8137
11. Mishra S, Tripathy HK, Mallick PK, Bhoi AK, Barsocchi P (2020) EAGA-MLP—an enhanced and adaptive hybrid classification model for diabetes diagnosis. Sensors 20(14):4036
12. Lemos R. Freelance writer—"3 cutting-edge data security technologies that will help secure the future". https://techbeacon.com/security/3-cutting-edge-data-security-technologies-will-help secure-future
13. Reuters T. Practical law. https://uk.practicallaw.thomsonreuters.com/w-017-1623?transitionType=Default&contextData=(sc.Default)&firstPage=true&bhcp=1. Access on 12-05-2020 @ 19:31 Malaysian Time
14. Rath M, Mishra S. Security approaches in machine learning for satellite communication. In: Machine learning and data mining in aerospace technology. Springer, Cham, pp 189–204
15. Mishra S, Sahoo S, Mishra BK (2019) Addressing security issues and standards in Internet of things. In: Emerging trends and applications in cognitive computing, pp 224–257. IGI Global
16. Petters J. Data privacy guide: definitions, explanations and legislation. https://www.varonis.com/blog/data-privacy/. Access on 12–05–2020 @ 19:31 Malaysian Time
17. M. Small. Kuppinger Cole analyst whitepaper—big data analytics—security and compliance challenges in 2019, Report No.: 80072. https://www.comforte.com/resources-detail/news/big-data-analytics-security-and-compliance-challenges-in-2019/. Access on 17th May, 2020

18. Rath M, Mishra S (2019) Advanced-level security in network and real-time applications using machine learning approaches. In: Machine learning and cognitive science applications in cyber security, pp 84–104. IGI Global
19. Mishra S, Tripathy N, Mishra BK, Mahanty C (2019) Analysis of security issues in cloud environment. Security designs for the cloud, Iot, and social networking, pp 19–41
20. Velumadhava Rao R, Selvamani K (2015) Data security challenges and its solutions in cloud computing. https://creativecommons.org/licenses/by-nc-nd/4.0/. Published by Elsevier B.V. Procedia Computer Science 48 (2015), pp 204–209, 205
21. Data Privacy & Security in Cloud Computing. https://www.apogaeis.com/blog/data-privacy-security-in-cloud-computing/. Access on 18th May 2020
22. Mishra S, Mallick PK, Jena L, Chae GS (2020) Optimization of skewed data using sampling-based preprocessing approach. Front Public Health 8:274. https://doi.org/10.3389/fpubh.2020.00274
23. Dutta A, Misra C, Barik RK, Mishra S (2021) Enhancing mist assisted cloud computing toward secure and scalable architecture for smart healthcare. In: Hura G, Singh A, Siong Hoe L (eds) Advances in communication and computational technology. Lecture Notes in Electrical Engineering, vol 668. Springer, Singapore. https://doi.org/10.1007/978-981-15-5341-7_116
24. https://docubank.expert/blog/5-best-data-security-technologies-right-now. Access on 23rd May, 2020
25. https://support.sas.com/documentation/onlinedoc/91pdf/sasdoc_913/base_datasecref_8946.pdf. Access on 23rd May, 2020
26. Mishra S, Mishra BK, Tripathy HK, Dutta A (2020) Analysis of the role and scope of big data analytics with IoT in health care domain. In: Handbook of data science approaches for biomedical engineering, pp 1–23. Academic Press
27. Information Technology Laboratory. Computer Security Resource Center. "FISMA Implementation Project". https://csrc.nist.gov/projects/risk-management/detailed-overview. Access on 23rd MAY, 2020
28. ISO/IEC 27001:2013. Information technology—security techniques—information security management systems—requirements. https://www.iso.org/standard/54534.html. Access on 23rd May, 2020
29. Mohapatra SK, Nayak P, Mishra S, Bisoy SK (2019) Green computing: a step towards eco-friendly computing. In: Emerging trends and applications in cognitive computing, pp 124–149. IGI Global
30. Mishra S, Koner D, Jena L, Ranjan P (2019) Leaves shape categorization using convolution neural network model. In: Intelligent and cloud computing. Springer, Singapore, pp 375–383
31. Buckley J (2020) https://www.qubole.com/blog/hadoop-security-issues/, October 15th, 2019, Access on 27th May, 2020
32. Jena KC, Mishra S, Sahoo S, Mishra BK (2017) Principles, techniques and evaluation of recommendation systems. In: 2017 International Conference on Inventive Systems and Control (ICISC), pp 1–6, IEEE

Printed in the United States
by Baker & Taylor Publisher Services